Building Responsible AI Algorithms

A Framework for Transparency, Fairness, Safety, Privacy, and Robustness

Toju Duke

Apress®

Building Responsible AI Algorithms: A Framework for Transparency,
Fairness, Safety, Privacy, and Robustness

Toju Duke
London, UK

ISBN-13 (pbk): 978-1-4842-9305-8
https://doi.org/10.1007/978-1-4842-9306-5

ISBN-13 (electronic): 978-1-4842-9306-5

Copyright © 2023 by Toju Duke

Managing Director, Apress Media LLC: Welmoed Spahr
Acquisitions Editor: Jonathan Gennick
Development Editor: Laura Berendson
Editorial Assistant: Shaul Elson
Copy Editor: Kezia Endsley

Cover image by Valery Fedotov on Unsplash (www.unsplash.com)

Distributed to the book trade worldwide by Springer Science+Business Media LLC, 1 New York Plaza, Suite 4600, New York, NY 10004. Phone 1-800-SPRINGER, fax (201) 348-4505, e-mail orders-ny@springer-sbm.com, or visit www.springeronline.com. Apress Media, LLC is a California LLC and the sole member (owner) is Springer Science + Business Media Finance Inc (SSBM Finance Inc). SSBM Finance Inc is a **Delaware** corporation.

For information on translations, please e-mail booktranslations@springernature.com; for reprint, paperback, or audio rights, please e-mail bookpermissions@springernature.com.

Apress titles may be purchased in bulk for academic, corporate, or promotional use. eBook versions and licenses are also available for most titles. For more information, reference our Print and eBook Bulk Sales web page at http://www.apress.com/bulk-sales.

Any source code or other supplementary material referenced by the author in this book is available to readers on GitHub (https://github.com/Apress). For more detailed information, please visit http://www.apress.com/source-code.

Printed on acid-free paper

To Emma and Alex, for teaching me what it means to be human and responsible.

Table of Contents

About the Author

 **Toju Duke** with over 18 years experience spanning across Advertising, Retail, Not-For Profit and Tech, Toju is a popular speaker, author, thought leader and consultant on Responsible AI. Toju spent 10 years at Google where she spent the last couple of years as a Programme Manager on Responsible AI leading various Responsible AI programmes across Google's product and research teams with a primary focus on large-scale models and Responsible AI processes. Prior to her time spent on Google's research team, Toju was the EMEA product lead for Google Travel and worked as a specialist across a couple of Google's advertising products during her tenure. She is also the founder of Diverse AI, a community interest organisation with a mission to support and champion underrepresented groups to build a diverse and inclusive AI future. She provides consultation and advice on Responsible AI practices worldwide.

About the Technical Reviewer

 **Maris Sekar** is a professional computer engineer, senior data scientist (Data Science Council of America), and certified information systems auditor (ISACA). He has a passion for using storytelling to communicate about high-risk items in an organization to enable better decision making and drive operational efficiencies. He has cross-functional work experience in various domains, including risk management, oil and gas, and utilities. Maris has led many initiatives for organizations, such as PricewaterhouseCoopers LLP, Shell Canada Ltd., and TC Energy. Maris' love for data has motivated him to win awards, write articles, and publish papers on applied machine learning and data science.

Introduction

I've always been a huge fan of technology and innovation and a great admirer of scientists and inventors who pushed the boundaries of innovation, some trying 99 times, 10,000 times, and more before achieving their goals and making history. Take the famous inventor of the lightbulb, Thomas Edison, or the brilliant Grace Hopper, who invented the computer. And before computers were transformed into machines, we had human computers, such as the super-intelligent "hidden figures," Mary Jackson, Katherine Johnson, and Dorothy Vaughan of NASA (National Aeronautics and Space Administration). We also have amazing geniuses like Albert Einstein, whose theories on relativity introduced many new ways of evaluating energy, time, space, gravity, and matter. Or the likes of Alexander Graham Bell, who introduced the telephone, and Josephine Cochrane, who we should thank for saving us from washing dishes by hand and invented the ubiquitous dishwasher!

These are just a few innovators and inventors who contributed greatly to technology, made our lives better and easier, and shed light on unknown phenomena. And there are many other sung and unsung heroes who contributed greatly to the world of science and technology.

Fast forward to today, to an ever-changing and evolving world: Humans are still inventing, creating, and introducing breakthroughs, especially in the field of technology. Many recent inventions are driven by artificial intelligence (AI), which is made up of deep learning networks (a form of AI based on neural networks, designed to mimic neurons in the human brain). For example, ChatGPT (a conversational AI built on a large language model), which is designed to provide intelligent answers to questions and solve many difficult tasks, has become the world's fastest growing app, with over 100 million users in just months. It still blows my mind how "intelligent" this app, and other dialogue AI systems similar to it, is. Another example is the various image recognition AI systems, which are used across the healthcare, automotive, criminal justice, agriculture, and telecommunications industries. We also have voice assistants such as Siri, Google Assistant, and Alexa, speech recognition systems, and so on. There's also DeepMind's (a UK-based AI company) Alphafold, which predicts 3D models of protein structures, contributing immensely to the medical field and driving further drug development and discovery. Alphafold solved a long-standing problem in the history of biology and medical science.

While we have these and so many more amazing use cases and success stories of AI and machine learning (ML) applications/products, it's important to note that there are also fundamental issues that plague these technologies. These range from bias, toxicity, harm, hate speech, misinformation, privacy, human rights violations, and sometimes the loss of life, to mention a few. Although AI technologies are great and highly beneficial to society in various ways, AI sometimes produces harm due to the lack of representative and diverse training data, lack of data ethics and curation best practices, less than optimal fine-tuning and training methods, and the sometimes harmful ways these AI applications are used.

In this book, I cover some examples where AI has gone drastically wrong and affected people's lives in ways that had a ripple effect on various groups and communities. Despite these various downfalls, I believe that AI has the potential to solve some of the world's biggest problems, and it is being used in various ways to tackle long-standing issues like climate change, as an example, by a good number of organizations. While we have many well-meaning individuals developing these highly "intelligent" machines, it's important to understand the various challenges faced by these systems and humanity at large and explore the possible ways to address, resolve, and combat these problems.

When tackling these issues across different dimensions, there's a potential framework for responsible AI that's worth adopting. This begins with understanding the gravity and importance of being responsible and accountable for models, products, and applications that could harm many people and communities, whether psychologically, emotionally, or physically; drive further inequality, leading to more poverty; violate human rights and privacy, and so on. A responsible AI framework can be built from the definition of principles that guide AI development. These include AI principles that address issues with data, incorporate fairness and its metrics, consider AI safety, take into account privacy, and build robust ML models.

In this book, I propose a simple, fundamental responsible AI framework that can lead to the development and deployment of ML technologies that have the safety and well-being of its end users, people, and customers in mind. Implementing further research on these different areas covered in this book can lead to less harm and bias across these technologies.

Understandably, it's difficult to measure the harm and damage that has been caused by these systems; it's far easier, on the other hand, to measure instances where AI has been beneficial and adopted across several industries. But given the challenges with this incredibly impressive technology, and the rapid rate of adoption in recent times, it's critical that we understand the benefits of AI as well as its limitations, risks, and challenges. My hope is that this book sheds some light on the incredibly positive potential of AI, but also on its limitations and shortcomings. The book guides you through potential ways to address these challenges, keeping humanity, users, consumers, and society in mind. I'll end with this note: being an adult is great, but with adulthood, comes great responsibility. This responsibility should be considered in the development of technologies that will not only be used and adopted by billions of people across the world, but that could potentially pose risks to these same people in many different ways.

PART I

Foundation

CHAPTER 1

Responsibility

The term *responsibility* is a relatable and simple term. Everyone, or almost everyone, deems themselves to be responsible in most if not every area of their lives. There's a sense of fulfillment and gratification when you think you have carried out a responsible act. Being *responsible* refers to carrying out a duty or job that you're charged with.[1] Most people in positions of authority feel a sense of responsibility to execute their jobs effectively. This includes parents, lawyers, law enforcement officers, healthcare professionals, members of a jury, employees, employers, and every member of society who has reached decision-making age. This chapter delves into AI responsibility and the need for building responsible AI systems.

Despite the fact that we were encouraged to be responsible at a very early age, it's often not accounted for in technology fields, and in particular in machine learning (ML) subfields, such as natural language processing, computer vision, deep learning, neural networks, and so on. Now you might argue that this isn't entirely true. There are ethical artificial intelligence (AI) and responsible AI practices developed every day. Although ethics and responsibility have been long-standing conversations that have taken place over the years, it's only recently, within the last 2-5 years, that we've seen an uptick in the adoption of responsible AI practices across industries and research communities. We have also seen more interest, from policy makers and regulatory bodies, in ensuring that AI is human-centric and trustworthy.[2]

© Toju Duke 2023
T. Duke, *Building Responsible AI Algorithms*,
https://doi.org/10.1007/978-1-4842-9306-5_1

There are several reasons that responsible AI has been slowly adopted over the years, the most prominent being that it's a new field that's slowly gaining recognition across the various AI practitioners. It's a bit sad that responsible and ethical practices were not adopted at scale, despite the 66 plus years of AI introduction.[3] Taking some cues from mental health experts, let's look at a few recommendations for acting responsibly.

Avoiding the Blame Game

Did you ever do something wrong with a sibling or a friend when you were a kid, and then tried to blame your accomplice once you were caught? My kids do this to me all the time. "I didn't do it; she did!" And the younger, not-so-innocent four year old, retorts, "No! It wasn't me! She did it!" Many people believe it's an inherent part of human nature to blame someone else when things go wrong. This could be a result of learned behavior and a desire to avoid punishment. It might also stem from a desire to seek parental approval. Research shows that adults who are quicker to blame others are more likely to have experienced some form of trauma in their lives that they haven't dealt with.

Before we go down the rabbit hole to the causes of the "blame culture," it's safe to say that people are often quick to blame someone/something else when things go wrong. This is something people need to unlearn in order to be and act responsible. When we have a case involving harmful ML models or products, the first question we tend to ask is "who's to blame?"

In site reliability engineering (SRE), it's a known fact that failure is part of the development process and is likely to happen quite often. A key part of the engineering process is to have a *post-mortem*, which is a written document that provides details of the incident, the magnitude of the incident, and the actions taken to resolve it, including its root cause. The document also includes follow-up actions/suggestions to ensure the problem doesn't happen again.

The problem with the blame culture is that it tends to negatively focus on people, which consequently prevents the right lessons to be learned—what caused the problem and how to prevent it from happening again. In light of this, a blameless post-mortem helps engineering teams understand the reasons an incident occurred, without blaming anyone or any team in particular. This, in turn, enables the teams to focus on the solution rather than the problem. The key focus is to understand the measures to put in place to prevent similar incidents from happening.[4]

If you've been working in the AI field long enough, particularly ethical AI, you've heard of the infamous "trolley problem," a series of experiments in ethics and psychology made up of ethical dilemmas. This problem asks whether you would sacrifice one person to save a larger number of people.[5] In 2014, an experiment called the Moral Machine was developed by researchers at the MIT Media Lab. The Moral Machine was designed to crowdsource people's decisions about how self-driving cars should prioritize lives in different variations of the trolley problem.[6]

In a (paraphrased) scenario where a self-driving car's brakes fail and it has two passengers onboard when approaching a zebra crossing with five pedestrians walking across—an elderly couple, a dog, a toddler, and a young woman—who should the car hit? Should the car hit the elderly couple and avoid the other pedestrians, or should it hit the little boy? Or should the car swerve and avoid the pedestrians but potentially harm the passengers? In other words, should it take action or stay on course? Which lives matter more—humans versus pets, passengers versus pedestrians, young versus old, fit over sickly, higher social status versus lower, cisgender versus non-binary?

In cases in which AI-related failures led to injury or death, I believe everyone who was involved in the development of the offending vehicle should be held accountable. That is, from the research scientist, to the engineer, to the CTO, to the legal officer, to marketing, public relations, and so on. It's the responsibility of everyone involved in the development of

ML technologies, regardless of their role, to think responsibly during the development of the models and products and to think about the impact it could have on the end users, the consumers, and society.

Being Accountable

When people accept accountability, it means they understand their contribution to a given situation. Being accountable also means avoiding the same mistakes over and over again. In some cases, this requires giving an account or statement about the part the person had to play. Not so surprisingly, accountability is a key component of responsible AI. According to the Organisation for Economic Cooperation and Development (OECD), companies and individuals developing, deploying, and operating AI systems should be held accountable for their proper functioning in line with the OECD's values-based principle of AI.[7] Chapter 2 delves further into the topic of responsible AI principles.

Let's look at a couple of examples where AI drastically impacted the lives of certain members of society, and the responsible companies were held accountable. Before delving into these stories, I'd like to take a pit stop and state that I'm a huge advocate and supporter of AI. I strongly believe that AI has the potential to solve some of the world's most challenging problems, ranging from climate change, to healthcare issues, to education, and so on. AI has also been adopted in several projects for good, otherwise known as AI for Social Good.

For example, top tech companies such as Google, Microsoft, IBM, and Intel are working on projects ranging from environmental protection to humanitarian causes, to cancer diagnostics and treatment, and wildlife conversation challenges, among others.[8]

AI also has many business benefits, including reducing operational costs, increasing efficiency, growing revenue, and improving customer experiences. The global AI market size was valued at $93.5 billion in 2021, and it is projected to expand at a compound annual growth rate (CAGR) of 38.1 percent between 2022 to 2030.[9] This shows that AI has great potential to improve businesses and people's lives while tackling major, long-standing problems the world faces.

Now that you've learned that AI is a much needed technology, let's consider the story of Williams Isaac, a young man from Michigan who received a call from the Detroit Police Department to report to a police station while working in his office at an automotive supply company on a Thursday afternoon. Thinking it was a prank, he ignored the summons and drove home. Upon driving into his driveway, a police car pulled up and blocked him in. Williams was handcuffed and arrested in front of his wife and two young daughters. When his wife, Melissa, asked where he was being taken, she was told by one of the officers to "Google it."

Mr. Williams was driven to a detention center, where he had his mugshot, fingerprints, and DNA taken; he was held overnight. Upon interrogation the next day, after a series of conversations with two detectives and a surveillance video shown of the criminal in question, they discovered that they had the wrong man, due to a flawed match from a facial recognition algorithm.

Although facial recognition systems have been used by police forces across the world for more than 20 years, recent studies have found that facial recognition works best on Caucasian men. The results are less accurate for other demographics, mainly due to the lack of diversity in the images gained from the underlying datasets.[10] Facial recognition systems are quite flawed, and people of all backgrounds have been falsely arrested,[11] although this more greatly affects people of color.

Let's take a step back and think about the effect this false arrest could have had on Williams, his wife, and his two little girls, who watched their dad get arrested right before their eyes. Putting this into context, the psychological effects of the night spent in jail could have been grave. There are also potential emotional and physical effects. I can only imagine the confusion and anger Williams must have felt that night, wondering what could have led to his false arrest. Knowing that the mere fact that he's a black man living in the United States makes him an easy target for the police, Williams may not have been altogether surprised, but he must have been quite saddened and anxious, hoping he'd return to his family in good time.

In a world where racism and discrimination still very much exists, it's quite appalling to see these societal issues prevalent in technologies employed and used by people in authority. These people of authority are the same individuals who are employed to protect our communities. If they decide to use technology and AI systems in their jobs, it's their duty to ensure that these systems promote fairness and equal treatment of people from different backgrounds, cultures, ethnicities, races, disabilities, social classes, and socio-economic statuses.

The law enforcement agency that committed the blunder apologized in a statement,[9] stating that Williams could have the case and fingerprint data expunged. When we consider "accountability" and what it entails, an apology and removal of his record is not enough. What the county needs to do to is make sure this sort of life-changing error doesn't happen again. They need to aim for clean data, run tests across the different subgroups for potential biases, and maintain transparency and explainability by documenting information on the data and the model, including explanations of the tasks and objectives the model was designed for. Carrying out accuracy and error checks will also help ensure results are accurate and less biased.

Some facial recognition software has been banned in certain use cases in the United States, including in New York, which passed a law in 2021 prohibiting facial recognition at schools, and in California, which passed a law that banned law enforcement from using facial recognition on their body cameras. Maryland also passed a law that prohibits the use of facial recognition during interviews without signed consent.[12] Despite this progress, there has been a steady increase in states recalling their bans on facial recognition; for example, Virginia recently eliminated its prohibition on police use of facial recognition, only one year after approving the ban.[13]

I'm happy to state that a few tech companies—Google and most recently Microsoft, Amazon, and IBM—stopped selling facial recognition technology software to police departments and have called for federal regulation of the technology.[14]

Across the globe, there are only two countries in the world that have banned the use of facial recognition—Belgium and Luxembourg.[15] In Europe, the draft EU AI act released in April 2021 aims to limit the use of biometric identification systems, including facial recognition.[16] While some parts of the world are still deliberating on how they'll use facial recognition software, it's encouraging to see there are countries, regulatory bodies, and organizations that recognize the dangers of facial recognition technologies and are ready to hold businesses accountable.

Eliminating Toxicity

Distancing yourself from people who exhibit toxic traits is advice that mental health practitioners provide to anyone seeking guidance about responsibility. By the same token, removing toxicity from ML models is a fundamental tenant of responsible AI. As datasets are built from the Internet, which certainly includes human data and biases, these datasets tend to have toxic terms, phrases, images, and ideas embedded in them. It's important to note that "toxicity" is contextual. What one person regards as toxic another might not, depending on their community, beliefs, experiences, and so on. In this context, toxic refers to the way the model is used; it's "toxic" when used in a harmful manner.

Since ML models and products need datasets to work, the inherent toxicity and biases prevalent in the datasets tend to creep into pretrained and trained ML models, eventually appearing in the final products or models. Let's take a look at a few examples of toxicity that crept into some recent NLP models.

Meta (formerly known as Facebook) recently launched BlenderBot 3, an output of its open-source Large Language Model (LLM) OPT-175B[17] that was released in July, 2022. BlenderBot 3 is an AI chatbot that improved its conversational abilities by searching the Internet to learn about topics. It conversed with people on these topics, while relying on feedback for conversational skills and improvements.[18]

One of the main challenges people face with toxicity in ML models is due to its representation of human data taken from the Internet, from platforms such as Reddit, Twitter, Socrata, Facebook, Wikipedia,[19] and so on. If you've faced any form of online harassment, or witnessed it in action, you've had a front seat view of how toxic humans can be to other humans. If we're building conversational agents using Internet data, it's going to be flawed. The data will be flawed, skewed, full of biases, discrimination, toxicity, lies, and incorrect information. Within a few days of BlenderBot 3's release, this became clear.

BlenderBot 3 didn't have high regard for its owner Mark Zuckerberg. In a different conversation, it stated that Zuckerberg's company exploits people for money. I wish I could say that's the most toxic statement BlenderBot 3 made. It predictably moved on to racist rhetoric and conspiracy-based territories. Social media users saw conversations where the bot denied the results of the 2020 U.S. election, repeated disproven anti-vaxxer talking points, and even stated that the antisemitic conspiracy theory that Jewish people controlling the economy is "not implausible."[20] The bot was trained off conversations from the Internet and it mimicked some public views, many of which aren't favorable.

As George Fuechsel says, "garbage in, garbage out" (GIGO).[21] GIGO is an expression that says no matter how accurate a program's logic is, the results will be incorrect if the input is rubbish. BlenderBot 3 is a clear example of GIGO. Even so, BlenderBot 3 is considered a much more successful chatbot than the short-lived Microsoft Tay, which lasted for only 16 hours.

In early 2016, Tay was launched by Microsoft as an innocent AI chatbot. It was immediately launched into the world of Twitter after its release, with the aim to learn how to become a good bot. Microsoft described Tay as an experiment in "conversational understanding." Tay was designed to engage and entertain people through casual and playful conversation as they interacted online.[19] It took fewer than 24 hours for Twitter to corrupt Tay.[22] Not long after the bot was launched, people starting tweeting several misogynistic, racist, and Donald Trump remarks, which the bot quickly adopted.

Tay's Twitter conversations reinforced Godwin's law, which states that as an online discussion continues, the probability of a comparison involving the Nazis or Hitler will emerge. There were tweets where Tay was encouraged to repeat variations of "Hitler was right" and "9/11 was an inside job."[23] After roughly 16 hours of existence, Tay mysteriously disappeared from Twitter. Microsoft acted responsibly and prevented the once-so-innocent, but now turned "toxic parrot bot," from propagating further harm, although the reputational damage and harm was done.

Thinking Fairly

In August 2020, hundreds of students in the UK gathered in front of the Department for Education chanting "swear words" at the algorithm. Thousands of students in England and Wales had received their "A-level" exam grades, which were scored by an algorithm. Due to the pandemic and social distancing measures, the "A-level" exams were cancelled and the UK's Office of Qualifications and Examinations Regulation (Ofqual) decided to estimate the A-level grades using an algorithm.[24]

The historical grade distribution of schools from three previous years, the ranking of each student in their school, and previous exam results for a student per subject were used as input to determine scores.

Experts criticized the low accuracy of the algorithm and public outrage was on the algorithm's unfair results. For example, if a student from a given school didn't have the highest grades in the past three years (2017-2019), it was highly unlikely that any student from that school would get a high grade. Results showed that students from smaller schools, aka private schools, were more likely to benefit from grade inflation than those from larger, state schools. This means that students from poorer backgrounds and socioeconomic backgrounds were the victims of the inaccurate algorithm.

Within days of the protests, the officials retracted the grades. Once again, the lack of fairness and equal treatment in the algorithm led to public confusion and outrage.

Algorithmic fairness is another key component in responsible AI frameworks, and the algorithm developed by Ofqual failed on this front. Prior to the release of this algorithm, fairness tests should have been carried out to ensure all students were treated fairly and equally, with various tests conducted on the datasets. Thinking fairly is an act of responsibility and all practitioners working on ML models/AI systems need to adopt this way of thinking.

Protecting Human Privacy

Products and technologies must ensure that they do not infringe on human rights, but rather protect them. AI systems should aim to protect human citizens, maintaining the individual rights of people to exercise their thoughts and take action. Scientists in general understandably tend to get carried away and excited about the amazing accomplishments a research study uncovers, particularly if it's a groundbreaking piece of research. This of course applies to AI researchers as well. In some cases, this has resulted in lack of awareness of the end users and the impact the launched product or model might have on society, including unknown/unintended consequences. Quite unfortunately, it's usually people from minority groups, poor backgrounds, and poor socioeconomic statuses that bear the brunt.

Let's take a look at the example of the alleged Systeem Risico Indicatie (SyRI) case from the Netherlands. SyRI was introduced by the Dutch government as a digital welfare fraud detection system, designed to mitigate welfare fraud. It used personal data from different sources and allegedly uncovered fraud. In 2018, six organizations formed a coalition to sue the Netherlands government over SyRI. SyRI was found to be too opaque, lacked transparency, and gathered too much data where the purposes of the data was unclear.[25] SyRI also potentially had a bias toward people living in lower-income neighborhoods, as they were clearly the main ones seeking welfare. By 2020, a Dutch court decided the SyRI legislation was unlawful because it did not comply with the right to privacy under the European Convention of Human Rights.

This is the first of its kind, where policy makers were reprimanded by a judicial court on the use of an AI system, breaching the right to privacy. The judgement reminds everyone, especially policy makers, that fraud detection must happen in a way that respects data protection principles and the right to privacy. It's an act of responsibility to recognize and protect the human rights of people, while preserving their privacy.

Ensuring Safety

When developing algorithms, developers must ensure that AI is deployed in a safe manner that does not harm or endanger its users. AI safety is one of the dimensions of responsible AI and it's important to bear in mind when developing ML systems. As an example, there are several online applications and websites that cater to underage users. Online safety is of paramount importance for such sites, considering the mental, emotional, and developmental effects harmful content can have on children. Online safety measures aim to protect children from viewing violent, sexual, and pornographic content, exposure to false information, or the promotion of harmful behaviors, including self-harm, anorexia, and suicide. Oversharing personal information and involvement in bullying or online harassment are some of the risks considered in child online safety.[26]

It's important that model applications and products protect users' personal identifiable information (PII) by employing privacy methods such as federated learning and differential privacy. These methods help anonymize user data and ensure a form of online safety.

Most people don't create ML algorithms or models with the intent of harming people, but since AI reflects society, its biases and discriminatory actions still pose a significant risk. More often than not, AI propagates harm in the output it produces. It's therefore an act of responsibility to develop ML applications with users in mind and employ methods that promote fairness, safety, privacy, and human rights preservation, and to ensure that the models are deemed safe, trustworthy, and fair.

Summary

This chapter laid the foundation for responsible AI by looking at the term "responsibility" and what it means to be responsible and accountable while protecting human rights, preserving user privacy, and ensuring human safety. You saw various examples, from the well-known ethical question of the "trolley problem" to several real-life examples of AI models that displayed "irresponsible" behavior, and the detrimental effects they've had. The next chapter looks at the next building block of a responsible AI framework—principles.

CHAPTER 2

AI Principles

The first chapter set the foundation for responsible AI frameworks, kicking off with responsibility and a few examples of AI and its ethical limitations. This chapter delves into "AI principles," which are fundamental components of building responsible AI systems.

Any organization developing and building AI systems should base these systems on a set of principles, otherwise known as AI principles or guidelines. AI principles are stepping stones for all types of AI development carried out across an organization. They are meant to be the foundation for AI systems and describe how they should be responsibly developed, trained, tested, and deployed.[30]

A good number of organizations and governing bodies have a defined set of AI principles that act as a guiding force for these organizations, and beyond. AI communities have seen a steady increase in AI principles/ guidelines over the past few years. While the design and outline of AI principles are fundamental to the development of AI principles, it's important that governance and implementation processes are put in place to execute these principles across an organization.

Most AI principles aim to develop ethical, responsible, safe, trustworthy, and transparent ML models centered on the following areas: fairness, robustness, privacy, security, trust, safety, human-centric AI and explainability, where explainability is comprised of transparency and accountability. The first section of this chapter looks at fairness, bias, and human-centered values and explores how these apply to AI principles.

© Toju Duke 2023
T. Duke, *Building Responsible AI Algorithms*,
https://doi.org/10.1007/978-1-4842-9306-5_2

Fairness, Bias, and Human-Centered Values

I remember giving a talk on responsible AI a few months ago, and during the question and answer (Q&A) session, someone mentioned the following (paraphrased), "I work as a professor in computer science and I'm tired of hearing the word *fairness*. It's been used too many times now, it's like a buzzword." I acknowledged his frustration but my response was how pleased I was to hear that "fairness" was being overused, because five to ten years ago, it wasn't even recognized in machine learning communities. If it's suddenly as popular as a Hollywood celebrity, I'm all for it!

As experts say, overcommunication is better than undercommunication, and in many cases the saying "What you don't know won't hurt you," is incorrect, especially in the field of AI. AI algorithms are full of potential harm, safety, bias, and privacy issues potentially affecting consumers and users. In this case, what you don't know can't only possibly hurt you, your career, and the reputation of your organization, it could possibly hurt millions of people across the world, and potentially future generations as well.

You're probably wondering what fairness is and what the hoo-ha is all about. Simply put, fairness, also known as *algorithmic fairness* in the context of machine learning, is regarded as providing impartial treatment to people, especially those of protected statuses.[31] What's fair can mean different things depending on the context and the people in question.

It can be a confusing concept and fairness as a term has different definitions across several disciplines. For example, fairness in law refers to the protection of people and groups of people from any acts of discrimination and mistreatment, with a focus on prohibiting certain behaviors and biases. Emphasis is on ensuring that decisions are not based on protected statuses or social group categories. In the social sciences, fairness is viewed in the context of social relationships, power dynamics, institutions, and markets. In quantitative fields such as math, computer science, statistics, and economics, fairness is considered a mathematical problem, while in philosophy, it's regarded as a concept that is morally right and denotes justice and equity.

Arvind Narayanan, a computer scientist and associate professor at Princeton University, identified 21 definitions of fairness.[32] How can fairness be considered fair if it has so many definitions? Chapter 5 looks at the different fairness definitions in more detail.

Let's take a look at a few organizations that have set AI principles in relation to fairness.

Google

Google defined its AI principles in 2018, including fairness and bias.[33] The company states that it will "avoid creating or reinforcing unfair bias" and acknowledges that the distinction between fair and unfair biases is quite difficult and differs across culture and societies. It aims to avoid unjust impact on people, especially those from various backgrounds and social groups such as race, ethnicity, gender, nationality, income, sexual orientation, ability (disability), and political or religious beliefs.[34]

The Organisation for Economic Cooperation and Development (OECD)

The OECD is made up of 37 governments with market-based economies working toward the development of policy standards with the aim to promote sustainable economic growth.[35] According to the OECD's AI principles, AI should be developed in consistency with human-centered values such as fundamental freedom, equality, fairness, rule of law, social justice, data protection and privacy, consumer rights, and commercial fairness.[36]

In addition to fairness, the OECD understands the risks and implications AI systems pose to human rights, and it pays close attention to human-centered values that could be infringed upon by AI technologies.

AI value alignment is not a new topic in AI. Its goal is to ensure that AI is properly aligned with human values and addresses the need to add moral values to artificial agents (systems). Although this might sound quite exciting, and you're probably thinking this might be the solution to ethical issues in AI, value-alignment also comes with its own set of challenges.

The challenge of value alignment is in two parts. The first is technical and focuses on formally encoding values or principles in artificial agents so their outputs are reliable.[37] As an example, recall the various chatbots discussed in Chapter 1, where Tay parroted all forms of toxicity and bias at online users? I'm quite sure it wasn't the intention of Tay's creators to launch an online bot exhibiting all sorts of hateful speech. It's quite unlikely Tay was built on a value aligned framework given it's immediate treachery in its responses. This is a clear example of how challenging value alignment can be, as the output and responses of the mentioned chatbots were quite unreliable.

The second challenge is the prevention of "reward-hacking," where the agent looks for ingenious ways to achieve its reward, even though the reward might differ from what was intended.

According to the OECD, since human-centered values can be infringed upon intentionally or unintentionally, using values-alignment can help ensure that AI systems' behaviors "protect and promote human rights and align with human-centered values." This principle also acknowledges the role of Human Rights Impacts Assessments (HRIA), codes of ethical conduct, quality labels, and certifications intended to promote human-centered values and fairness.

The Australian Government

The Australian government has a set of AI principles that includes human-centered values and fairness. The government opines that AI systems should respect human rights and freedom and enable diversity and the autonomy of individuals while protecting the environment throughout their lifecycle.

The democratic process must not be undermined by AI, and AI systems should not engage in actions that "threaten individual autonomy, like deception, unfair manipulation, unjustified surveillance, and failing to maintain alignment between a disclosed purpose and true action."[38]

The Australian government intends to ensure that their AI systems are fair and enable inclusion throughout their lifecycle. The systems should also be user-centric, involving appropriate consultation with stakeholders that might be affected and ensure people receive equitable access and treatment. They also suggest that AI produced decisions should comply with anti-discrimination laws, given the potential AI has to perpetuate social injustices and disparate impact on vulnerable and underrepresented groups.

Let's look at an example of fairness put into practice by Google. In 2015, a software engineer named Jacky Alciné noticed that the Google photos app was classifying him and his African-American friend as "gorillas" when he carried out a search on the app. He immediately went on Twitter to report his disappointment, which resulted in profuse apologies from Google and several press statements. As Google photos uses image recognition algorithms, Jacky's picture must have been mislabeled by the algorithm, which comes as no surprise, as image recognition systems have a huge propensity for mislabeling errors.[39]

To provide an immediate solution, Google removed all labels of gorillas and chimpanzees from the app. Although this does not solve the problem in its entirety, it's a good starting point and shows that Google is committed to its AI principles to not "reinforce unfair biases."

It also shows that fixing bias and fairness issues in AI systems is no small feat, and it's impossible to find an AI system that is 100 percent bias free today (who knows, this might be different in the future). At its best, there could be a significant reduction of biases, but not a total elimination, due to various factors such as data labeling, model development, and so on.

Building transparency and trust into these systems is fundamental and key to having responsible AI systems, and it helps determine fairness and bias issues. The next section covers this topic a bit more.

Transparency and Trust

In the United States, judges, probation officers, and parole officers used risk-assessment AI algorithms to assess a criminal defendant's likelihood of becoming a *recidivist*—a term used for criminals who reoffend.[40] In 2016, ProPublica, a nonprofit investigative agency based in New York, carried out an assessment of a tool called COMPAS (Correctional Offender Management Profiling for Alternative Sanctions) used by several states. They wanted to determine the underlying accuracy of this recidivism algorithm and see if the algorithm was biased against certain groups.

After carrying out the investigation on over 10,000 criminal defendants in Florida's Broward County, they discovered that "defendants from African American descendants were far more likely than Caucasian defendants to be incorrectly judged of recidivism, while Caucasian defendants were more likely to be incorrectly flagged as low risk compared to African American defendants." The results of the investigation led to the deprecation of the software in 2016.

Due to the well-known "black box" problem in AI, where it's quite difficult to understand the reasoning and methodologies applied to AI outputs, businesses and the society at large have had a growing distrust of the technology. Who can blame them, given the repeated errors discovered in these systems? This distrust in AI is probably its biggest obstacle to widespread adoption.[41] This led to the introduction of *explainable AI,* commonly known as *explainability,* which aims to solve the "black box" problem. Chapter 7 looks at this concept more closely.

On the upside, AI is increasingly helping organizations become more efficient and productive, while working toward crucial decisions such as loan disbursements, job placements, and medical diagnoses, among others. On the downside, an unreliable AI system that can't be trusted due to several factors, including ethics and responsibility, could lead to a loss of customers, profit, and reputation, including lawsuits and regulatory scrutiny, as is the case today with several organizations.

In order to build trust in AI, transparency and explainability measures need to be in place. Transparency helps explain why an AI system made the prediction it did. It contributes to repairing trust in these systems, including explanations of algorithms, uncertainty estimates, and performance metrics.[42]

Most organizations and governmental bodies that take responsible AI seriously ensure they have trust and transparency embedded in their AI principles. A good example is IBM. According to IBM, users must be able to see how the service works, evaluate its functionality, and comprehend its strengths and limitations as part of its transparency offerings. As an organization, they believe that transparency reinforces trust, and transparent AI systems should share information about what data was collected, how it will be used and stored, and who has access to it.[43]

Transparency provides information about the AI model or service, which helps consumers better understand the system and decide whether it is appropriate for their situation.

In a similar vein, the Alan Turing Institute published guidelines in 2019 on how to apply principles of AI ethics and safety to AI algorithms in the public sector. The institute defines the justification of transparent AI when the design and implementation processes that have gone into a particular decision of a system can be demonstrated and the decision or behavior itself is ethically permissible, fair, and worthy of public trust.[44]

Accountability

You've seen a good number of examples where AI has gone wrong, produced incorrect and false results, spat out misogyny and discrimination, and falsely accused people of crimes they didn't commit. This books has yet to review more tragic examples, where autonomous vehicles such as self-driving cars have killed pedestrians, or wrong medical diagnoses led to a higher fatal rate in breast cancer among African American women compared to Caucasians. The main question in all of this is who should be held accountable when AI goes wrong?

Accountability refers to the need to explain and justify one's decisions and actions to partners, users, and any other person the AI system interacts with. In order to ensure accountability, decisions must be explained by the decision-making algorithm used during the process, including the representation of moral values and societal norms the AI agent uses during the decision-making process.[45]

An example of accountability requirements that were properly outlined is found in the U.S. Government's Accountability Office (GAO), which developed the first framework for the federal government to ensure the accountable and responsible use of AI systems. The framework discusses the inclusion of accountability in the entire AI lifecycle, from design and development to deployment and monitoring. It lays out specific questions for evaluating and auditing procedures along four dimensions: governance, data, performance, and monitoring.[46] Chapter 8 looks at these dimensions more closely.

Circling back to the OECD, this organization also has accountability as one of its AI principles and defines accountability "as an ethical, moral, or other expectation that guides individuals' or organizations' actions and allows them to explain reasons for which decisions and actions were taken." It references liability and the adverse legal implications that could arise from an individual's or organization's action or inaction.[47]

The trolley problem covered in Chapter 1 sheds some light on accountability as an AI principle. The question on who should be held liable and accountable if a self-driving car causes harm to passengers or pedestrians shows the need for including ethical values throughout the ML lifecycle of an AI system.

Social Benefits

The humongous potential of AI, including its ability to solve and tackle some of the world's most challenging social problems, was explained in the first chapter. There are several projects that have taken place and are still ongoing that contribute to the UN's sustainable-development goals, potentially helping hundreds of millions of people in advanced and emerging countries.

In a study conducted in 2018, McKinsey observed that AI use cases for social good were included in the following domains:[48]

- Security and justice
- Crisis response
- Economic empowerment
- Education
- Environment
- Equality and inclusion
- Health and hunger
- Information verification and validation
- Infrastructure
- Public and social sector

Among these domains, the following AI capabilities were mapped to potential uses that will benefit society:

- Deep learning algorithms

- Natural language processing

- Image and video classification

- Object detection and localization

- Language understanding

- Audio detection

- Sentiment analysis

- Language translation

- Face detection

- Tracking

- Emotion recognition

- Person identification

- Optical-character and handwriting recognition

- Speech to text

- Content generation

- Reinforcement learning

- Analytics and optimization

While some of the functionalities outlined here are questionable and have raised a few eyebrows and red flags from AI research communities (such as emotion recognition and face detection), most of these functionalities can help build socially beneficial AI systems.

Deep-learning algorithms, a part of AI based on neural networks, have been heavily adopted in healthcare settings, especially imaging capabilities, which are used for cancer identification and screening. AI is also being utilized to predict the development of diseases.

Mount Sinai, a New York-based healthcare organization, used deep learning to predict the development of liver, rectum, and prostate cancers, achieving 94 percent accuracy.[49] It's also been reported that AI contributed to the development of the COVID vaccine.[50]

Due to the amount of published cancer research, clinical trials, and drug development, there is a huge amount of data that AI systems can use to guide decisions in healthcare. How representative this data is of minority and underrepresented groups is still in question.

In 2021, a UK based AI company called DeepMind developed a method to "accurately predict the structure of folded proteins and mapped 98.5 percent of the proteins used within the human body." The prediction of the structure of nearly all proteins known to science in 18 months is a transformational breakthrough that will speed drug development and transform basic science.[51] DeepMind developed an AI system called AlphaFold, which predicts a protein's 3D structure from its amino acid sequence.[52] This problem has existed for decades, as it's been difficult for scientists and researchers to determine the crumpled and folded shapes of proteins based on their sequences of amino acids.

This discovery has already led to progress in combating malaria and antibiotic resistance.

Alongside healthcare breakthroughs, AI is being used across several other projects, like the "save the bees" campaign, where the study of bee survival is being analyzed by AI models to identify patterns and trends. These patterns could lead to early interventions for bee survival, which are important for the planet and food supply. AI is also being used to create tools for people with disabilities, tackle climate change, conserve wildlife, combat world hunger, and so on.

AI used for societal benefit spans across many other domains, including agriculture. As the United Nations (UN) forecasts that the world population will grow to 9.7 billion by 2050,[53] there will be a massive strain on food production to ensure food security and availability across the world's population. To attain food security, nutritious and safe food has to be available to everyone. Food security and global poverty is another challenge facing humanity. According to the International Food Policy Research Institute (IFPRI), there are about 795 million people facing hunger every day and more than 2 billion people lacking vital nutrients, such as iron, zinc, vitamin A, and so on.[54]

With a steadily growing population and a high demand on food production due to the increasing number of people to feed, climate change further compounds the situation. The nutritional value of food, such as grains, legumes, and tubers is affected by the elevated levels of CO_2. To put this into perspective, according to the U.K's government program on food security and research, "Nearly a quarter of all children aged under five today are stunted, with diminished physical and mental capacities, and less than a third of all young infants in 60 low- and middle-income countries meet the minimum dietary diversity standards needed for growth."

It's been estimated that we'd need to produce more food than has ever been produced in human history in the next 35 years due to the increased projections of the world's population and the need to change diets due to rising incomes. Food production will need to increase by 60 percent in order to feed an additional 2 billion people. This problem intensifies with these additional challenges:

- No new land for agriculture (with a projection of only 4 percent increase in land by 2050)

- Increasing sea levels, which invariably reduce land availability and the need for land to produce bioenergy

- Carbon capture and storage (BECCS), which
  helps remove greenhouse gases (GHGs) from the
  atmosphere, a solution toward reducing the effects of
  climate change

In other words, the demand for land is high, and increasing focus on land for agricultural purposes is being reduced due to climate change. This means we need to produce more food without further expansion of agricultural areas, also known as sustainable intensification (SI) of agriculture on land.[55]

In 2021, the High-Level Political Forum on Sustainable Development (HLPF) carried out a review of the United Nations (UN) Sustainable Development Goal (SDG) 2 on zero hunger, and in its findings, it stated the world was already "off track in achieving zero hunger by 2030 and healthy diets were inaccessible for a significant part of the world's population." With the most recent COVID-19 pandemic, it is estimated that the pandemic further increased the "number of people suffering from chronic hunger from 83 million to 132 million people in 2020, further adding to the 690 million estimate in 2019."[56]

These are real issues that need viable solutions, and guess what? AI can come to the rescue. AI is increasingly helping to solve food production and agriculture problems in many ways. According to the Food and Agriculture Organization of the United Nations (FAO), these methods can be categorized into three parts: 1) agricultural robotics, 2) soil and crop monitoring, and 3) predictive analytics.[57]

Let's take a look at a few examples where AI is helping solve one of the world's top problems—hunger—using agriculture. In 2017, a team of researchers used Google's TensorFlow (a ML/AI open source library) to build a library of images for cassava leaves from cassava plants (one of the most widely grown root crops in the world) in Tanzania, which helped identify disease in these plants with 98 percent accuracy. Using transfer learning (a ML technique that repurposes a model on a different task from the one it was initially trained on), the researchers built over 2,000 images of cassava leaves and then trained the model to recognize common diseases and pests found in the plant, such as brown leaf spot disease (a fungal disease that affects leaves) and the effects of red mites.[58]

Using Meta's PyTorch platform—an open source deep learning ML platform—two graduate students from Stanford (not Larry and Sergey this time!) developed a "See and Spray" machine, which is a device that can be used by tractors to detect weeds in crop fields and spray the unwanted plants with herbicide, while avoiding the crops. Leveraging computer vision using its front and rear cameras, the machine uses ML to determine if plants are crops or weeds, processing plant images 20 times per second. It then compares the processed images to its training library containing over one million images. This incredibly useful technology enables farmers at scale to manage large fields containing millions of plants, thereby reducing their herbicide costs. The machine only targets weeds, versus spraying the entire field, which translates to huge amounts of cost savings for the farmer, and indirectly helps improve food security, because less food is wasted during farming. "See and Spray" also provides insights on data analytics, whereby farmers can determine the types of herbs, their location, and the total number of herbs in their fields.[59]

Tackling climate change, AI also offers several solutions to address this long-lasting challenge facing our planet. These solutions range from robots and drones powered by AI to help with wildfires, which have become more rampant and pose an increasing threat to human lives, communities, landmarks, and nature reserves, while leading to further global warming. To this effect, AI-powered robots have been introduced to help with firefighting. As firefighters have to constantly deal with toxic smoke, collapsing structures, and so on, autonomous firefighting robots are being employed to predict and locate fires as well as put them out and collect hazardous data. Concurrently, drones are being utilized to provide surveillance to firefighters without the need for human intervention. In some cases, the drones also provide a thermal video functionality, which helps firefighters do their job more effectively while reducing water consumption, leading to a more sustainable service. Fire service departments such as the Los Angeles City Fire Department (LAFD) have employed the use of AI-powered robots and drones for their operations.[60]

AI is combatting climate change and global warming in several other areas, including "weather prediction, energy efficiency, or by reducing emissions from transportation, agriculture, and industry" as stated by the AI for Good organization.[61] For example, green-cooling technologies cool electronics without the use of natural resources. Green-cooling technologies can "reduce energy consumption at data centers from 20 percent to 4 percent if applied at scale."

Another challenge that contributes to climate change is plastic pollution, a threat to human health and marine ecosystems. Plastic waste has led to the suffocation and death of many marine species; microplastics infiltrate tap water, beer, and salt, and are known to be carcinogenic and cause "developmental, reproductive, neurological, and immune disorders in both humans and wildlife." In addition, plastic pollution affects climate change, as it releases "carbon dioxide and methane (from landfills) into the atmosphere, thereby increasing emissions,"[62] as cited by the International Union for Conservation of Nature (IUCN).

To help combat the deleterious effects of plastic pollution on the world's oceans, an organization called Ocean Cleanup aims to "remove 90 precent of floating ocean plastic by 2040" using AI. Using remote sensing, their AI technology detects and maps ocean plastics. This detection process enables the organization to locate areas with high amounts of plastic debris, which they clean up once identified. As an example, they scanned an area of the ocean for marine debris using GoPro cameras and used the images to build the dataset that feeds the AI object detection algorithm. The exercise revealed over 400 large plastics in the images and helped them identify the locations of these debris, leading to further identification of affected areas and the potential removal of plastic waste.[63]

In addition to these numerous examples, there are also a few other ways that AI is being utilized to combat climate change, such as the development of smart cities. Countries including Singapore, Norway, Finland, Switzerland, and South Korea are making their cities more sustainable and energy efficient by using satellite imagery to predict deforestation.

Tech companies like Google, and Microsoft, as well as many other organizations including nonprofits, are utilizing AI to tackle prevalent social problems. For example, Google launched the Tree Canopy Lab in 2020, an "environmental insights" tool that uses AI and aerial imagery to provide insights to cities on their tree canopy coverage and assist cities with the measurement, planning, and reduction of carbon emissions.

As countries are thinking of different ways to tackle climate change, another potential solution is to increase the shade in cities due to the steady growth of "heat islands, which are areas that experience higher temperatures, leading to poor air quality, dehydration, and other public health concerns." Trees are one way to reduce this problem, as they help lower street level temperatures, overall improving the quality of life and contributing toward climate change mitigations and solutions.[64]

Google is also contributing to flood forecasting in India and Bangladesh, partnering with the Red Cross Society, India's central water commission, and the Bangladesh Water Development Board. They are using ML, physics-based modelling, and alert-targeting models to identify areas that are at risk of wide-scale flooding, using publicly available data. This effort has been applied to areas with over 360 million people and has sent over "115 million alerts," alerting areas at risk of flooding before they arise. This prevents the loss of human lives that may have been caused by floods.[65]

Microsoft's AI for Good projects spans many areas, ranging from health, climate change, accessibility, humanitarian action, and cultural heritage. Deep diving into Microsoft's humanitarian focus, the company has committed $40 million to support nonprofits and organizations working on humanitarian projects. The company strives to support "disaster response, refugees, displaced people, human rights, and women and children's needs."[66] One way they are working on AI for Good is using conversational AI through a chatbot to support women suffering from gender-based violence in Puerto Rico. Post-pandemic, an increase in domestic violence was observed toward women, and the country was declared a state of emergency due to gender-based violence in 2021. This project is in partnership with Seguro, a nonprofit organization.[67]

Another fundamental dimension of responsible AI that should be included in all governing AI principles is preserving user privacy and security while ensuring safety. This is covered as the next topic.

Privacy, Safety, and Security

The advancement of technologies, especially super intelligent ones like AI, should not lead to privacy violations, data breaches, and security. As mentioned, its primary use should be to benefit human citizens while enabling business and organizations' efficiency, productivity, and ultimately increased profit margins. This is the ideal situation for the use of AI, which makes it important to consider human safety, privacy, and security in all AI principles and systems, regardless of the country, organization/business, or regulatory body.

In UNESCO's recommendation on the ethics of AI, it outlines that unwanted harms (that is, safety risks and security risks) should not only be avoided, but should be addressed, prevented, and eliminated throughout the lifecycle of all AI systems.[68]

The agency also states that privacy, which involves the protection of human dignity, autonomy, and agency, must be "respected, protected, and promoted" throughout the AI lifecycle. Data used by AI systems should be collected, used, shared, archived, and deleted in ways that are consistent with international laws and should respect relevant national, regional, and international legal frameworks.

In April 2019, the European Commission (EC) published ethical guidelines for trustworthy AI, which includes guidance on privacy and data governance. Privacy is seen as a method to prevent harm, and requires adequate data governance that covers the quality and integrity of the data used, and its relevance with respect to the domain of the AI system(s) and where it will be deployed.

The EC's ethics guidelines also state that privacy and data protection should include user information provided over the course of the interaction with the AI system. Data collected from individuals must not be used to unlawfully or unfairly discriminate against them and digital records of human behaviors must not be tampered with by any AI system in regard to their individual preferences, sexual orientation, age, gender, and religious or political views.[69]

Companies must outline how processes and datasets are used at every step of the lifecycle, such as planning, training, testing, and deployment. There should be data protocols in place for governing access to data. These protocols should outline who can access the data and under what circumstances. Only qualified staff with the competence and need to access an individual's data should be allowed to do so.

It's important to note that there's a clear difference between the United States privacy laws and the European Union's (EU). In May 2018, the European Parliament and Council of the European Union introduced the General Data Protection Regulation (GDPR). This led to many organizations in the EU and European Economic Area (EEA) scrambling to comply with its laws, also facilitating the swift emergence of many startups aimed at assisting organizations with the new laws. If you're new to the game, GDPR kicked off the privacy and cookie policies you see on websites when you're browsing within the EU, or when you're visiting an "EU owned" site.

GDPR is the toughest privacy and security regulation in the world. It's designed to protect the data privacy of all people in the EU, and it issues harsh fines to any organization that violates its laws. These laws relate to the collection, storage, and possession of personal data. Meta, Google, and Clearview AI were among the companies that received the largest GDPR fines in 2022, ranging from $405 million to $6 million. Amazon still takes the trophy for being hit with the largest fine of $746 million in 2021.[71]

GDPR's rules are quite substantial and include the "minimization of data collection, storage limitation, and accountability," among others. Unfortunately, the United States doesn't have a similar comprehensive data privacy law that applies to all types of data. It's speculated that this could be due to the cultural differences between both regions, where the EU's privacy-first mindset stems from the history of National Socialism and Communism when people's information was used for heinous purposes. On the flip side, the United States seems to give more control and power to organizations collecting user data, as opposed to the individuals who supposedly own the data, leading to the question that never grows old—"who owns the data?" Is it the individual or the company? According to the EU and GDPR, it's definitely the individual. After all, the data is information about the person in question.

The United States' approach to privacy and data collection is more fragmented, where we see various industries left to decide how to enact privacy laws. To date, the state of California's privacy law is most comparable to the GDPR. It's called the California Consumer Privacy Act (CCPA) and it protects consumers residing in California by giving them more control over the personal information that businesses collect about them. It also shares ways on how to enact the law if there are any violations. The act has recently been updated to include opt-out rights and imposes additional regulations on businesses, taking it a step closer to alignment with GDPR.

Looking at some of the different privacy acts that exist across the various industries in the United States, there is the Health Insurance Portability and Accountability Act (HIPAA), which protects patient healthcare information by setting rules on how healthcare providers secure patient data against fraud and theft.

The Gramm-Leach-Bliley Act (GLBA) focuses on the finance sector, where it sets out responsibilities and standards to financial institutions with the goal of protecting and securing consumers' personal information. Another regulation is the Federal Information Security Management Act (FISMA), which requires federal agencies to develop information security programs, thus protecting against cyber threats to federal information systems.[70]

Let's look at example in which a company was penalized for violating data privacy laws. If you're a fan of good old AI (like my humble self) and all the shenanigans it faces on a daily basis, you have probably heard of Clearview AI Inc., a startup located in New York, which recently received a fine of approximately 7.5M GBP by the ICO (Information Commissioner's Office). It used images of people in the UK and elsewhere that were scrapped from online sources to create a global online database that could be used for facial recognition.

Clearview AI Inc., using the Internet, collected more than 20 billion images of people's faces and data from across the world to create an online database, without people's knowledge or consent. This data allowed its customers, including the police, to upload an image of a person to the company's app. It then listed images with similar characteristics and provided links to the images.[71] I'm still trying to figure out what sort of business model this is, but Clearview "clearly" has a lot of customers across the globe, so it must have potential?

I've seen online interviews with the founder of Clearview, who repeatedly mentioned that the company was legit and there was nothing wrong with their services or methods of collecting data. This is another good example of privacy violations and lack of responsibility/ethics.

While it's great to see regulatory bodies stepping up to protect its citizens from the likes of companies like Clearview, suffice it to say that Clearview's services still exist outside the UK, and it still has the world at its fingertips, literally.

Let's imagine a world without any guiding principles to support ML development. It'd be a very terrible world to live in. If you've used any of the recent AI chatbots for example, you've seen how temperamental and schizophrenic they can be, spewing out responses that threaten the user, sometimes providing glaringly incorrect responses. They sometimes argue endlessly with the user on how factual they are, including lots of hallucination that is borderline scary. Existing and upcoming regulation will potentially protect society from known harms, but it's up to the developers, engineers, and organizations developing these models/products to ensure they are ethical, responsible, and safe for everyone in the world, regardless of location, beliefs, gender, religion, and so on. Developing AI principles before developing ML programs is a good step in the right direction.

Summary

This chapter reviewed the first building block for developing responsible AI systems. It delved into various dimensions of responsible AI, including fairness, bias, human-centered values, privacy, security, safety, and the societal benefit of AI. It looked at various examples of AI principles from several organizations, including UNESCO, the European Parliament, OECD, Google, and Microsoft. The next chapter moves on to the next building block—data—where you read about its history, the ethics around it, and tools for accessing it.

CHAPTER 3

Data

After setting up and understanding the AI principles, the next foundational element and part of responsible AI development is looking at the data and its guiding principles.

Data is the underlying ingredient of all ML systems. As ML works with loads of data, it's critical to understand the foundations of data and data ethics, with the primary focus on training and fine-tuning your model appropriately.

The "80 Million Tiny Images" dataset was built by MIT in 2006 and it contains different nouns from Wordnet, a large database for words using semantic relations such as synonyms, et al. The dataset has over 1,700 citations according to Google Scholar.[72] Dublin-based researcher Abeba Birhane, and Lero, a SFI software research center, discovered that the dataset contained a high amount of racist insults, misogynistic slurs, and hate speech. These included references to the N-word, and slurs such as "whores," "rape suspect," and "child molester," among other offensive terms. It also had content like non-consensual photos of nudity and inappropriate images of children.

This is yet another example where linking images to offensive language further perpetuates prejudice and bias in AI and ML models, driving further stereotypes and prejudices of discriminated against groups in society.[73]

© Toju Duke 2023
T. Duke, *Building Responsible AI Algorithms*,
https://doi.org/10.1007/978-1-4842-9306-5_3

MIT immediately made a press statement and took the dataset down, taking it offline as a result of the research and negative press that followed. In the statement, they mentioned the infeasibility of investigating the images due to their small size of 32 x 32 pixels, making it difficult for people to view the content, and subsequently reducing the possibilities of manual inspection.[74]

The History of Data

Before delving further into the topic of data ethics, let's take a step back to look at the history of data and how it all came about. This history lesson is important, as it sheds light on the challenges data faces, and subsequently ML/AI models and systems also face. Gaining further insights into the underlying issues can help address these prevalent problematic challenges.

Data traces back to the Upper Paleolithic Period, as early as 19,000 BC where the Ishango bone, a fibula of an animal, was unearthed in the Belgian colony of the Congo (now the Democratic Republic of Congo).[75] It's presumed to have acted as a tally stick, representing a very early instance of data collection and storage. The scratches on the bone are assumed to be the first documented record of logging numerical information for later use, although this is still under debate.[76]

By the 1600s, data saw its first English use. Derived from Latin, data means "a fact given or granted." Another important data-related event took place in the late 1800s with the collection and analysis of the U.S. census data by the Hollerith machine, a machine invented by Herman Hollerith that processed and analyzed large amounts of data.[77]

The 1900s saw a shift to data storage and collection, which led to a series of events, from the invention of magnetic tape for recording processes, to the concept of cloud storage data, and the introduction of a framework of database management, which is still being used today.[78]

The well-known "Internet" was also introduced in the 1990s by Sir Tim Berners-Lee with the creation of hyperlinks and hypertext, taking data to the next noteworthy level of enabling worldwide access. By the mid 1990s, AT&T launched the first instance of all web-based storage, and in 1997 Google search was launched. The launch of Google search led to data access for everyone using computers, and, as the saying goes, "the rest is history."

You could say that data came about by happenstance, so it's not surprising that data ethics and the potential impact of data has been somewhat overlooked.

Data Ethics

The growing use and popularity of data in recent years has led to the increased awareness of the need for data ethics and guidelines on the collection, use, and storage of data.

The "80 Million Tiny Images" example mentioned earlier in the chapter highlights the moral problems existent in data. These problems exist in the curation, generation, documentation, processing, disseminating, and using of data. This includes algorithms and any corresponding practices.

Data ethics is considered a new branch of ethics that builds on the foundation of computer and information ethics, with a more refined approach to data-centricity. It highlights the various moral dimensions of data and the need for ethical analysis and evaluation. It encourages the good and responsible practice of collecting, sharing, and using data, with particular focus and attention on its potential to adversely impact people and society.[79]

The adoption of data ethics has risen over the years when Edward Snowden, a former computer intelligence consultant, leaked highly classified information from the NSA (National Security Agency) and

Facebook's interference in the U.S. presidential election in 2016, which led governments to publish and enforce data ethics litigations such as the European Union's General Data Protection Regulation (GDPR).

Most private and regulatory bodies, including governmental agencies, are coming together to provide data ethics guidelines, with the aim of increasing awareness, understanding, and adoption.

As with most ethical practices, principles are set in place to guide practitioners who seek further guidance on ethical/responsible AI processes. On that note, data ethics has a few guiding principles designed with the users in mind to help with the integration of data ethics in data-driven activities. The following sections discuss a few data ethics principles to consider.

Ownership

This is the first principle of data ethics. It states that the individual should have ownership of their personal information, where the collection of someone's personal data without their knowledge or consent is unlawful and unethical.[80]

Common ways to gain consent are through signed agreements, digital privacy policies, consent cookies on websites that track users online behavior, and so on. The blanketed assumption that a consumer or customer is fine with the collection of their data is incorrect, and if you speak with most people today, they are scared and frustrated at how their data is collected, where it's stored, what it's used for, and how much control or say they have over it.

The other day I was working away, and I saw an incoming call from a number I didn't recognize. I decided to pick up the call as it looked like a number that had attempted to contact me a few times. Here's how the conversation went:

Me: "Hello"?

Caller: "Hello. Are you interested in online trading…?"

Me: "No, I'm not but wondering where you got this number from?"

Caller: "Oh, you probably added it to a website, or an email, or something..."

Me: "Well I'm not sure where you got my number from, and I'm not comfortable or interested in..."

Caller: Hangs up the phone.

My concern with data privacy and consumer protection/rights increased after this very recent conversation. I'm still perplexed at how third-party agencies sell people's data without caring about how it will be used or to what extent companies might go to sabotage innocent people. Another good example are the scam texts that attempt to hack people's bank accounts. These are related to data, privacy, and lack of data ethics/principles.

We constantly hear of hacks where people's personal identifiable information (this refers to any information that could be used to identify an individual, including full name, date of birth, home address, phone number, social security number, national insurance number, and so on) is leaked. The only recourse the affected people can take (assuming they're even aware of the issue) is to change their passwords. This means they have no control over their personal information, which should not be the case. This leads to the next principle of data ethics, data control.

Data Control

Individuals should have primary control over their data and proper context as to how their data is processed and activated. They should be able to manage the flow of their personal information and be able to opt out at any time. As an example, a good number of countries have passed laws that provide consumers with an amount of control over their contact information through opting out of emails or other messages.[81] I must say, the "opt out" button is one of my favorites!

Transparency

Data ethics recommends including transparency in your data practices. All data-processing activities and automated decisions should be understandable and made known to the users. The purpose of data processing with respect to risks, and social, ethical, and societal consequences must be clear to the individual and be easy to understand. If there are any changes to the way the customer data is used, it's the organization's moral and professional obligation to inform the customers of those changes, and ultimately request their permission for the ongoing use of their data. Transparency is also required by many existing privacy laws across the globe.

Users also have the right to know how you plan to collect, store, and use their data. Many consumers do not have a clue whether their information is shared, and it's the responsibility of the organization to ensure that this information is easily understood in as simple terms as possible, so that all users—regardless of age, educational background, disability, and so on—can understand how their data will not only be used, but also stored. It's also important to provide information about how they can opt out if they desire to do so.

Accountability

Accountability is a fundamental part of data processing where efforts should be made to reduce the risks and mitigate social and ethical implications for individuals. Proper handling of data processing should be embedded throughout the organization and accountability should be included as part of the data workflow. This could be in relation to forecasts and analysis, where there must be fair representation of consumers across the datasets in question. In other words, you don't want to be that "guy" or organization furthering discrimination and unfair biases through the datasets used in your AI systems. Like MIT, mentioned at the beginning of this chapter, which recalled their datasets after biases where discovered.

Equality

You may have heard the phrase "democratizing AI," which basically means making AI more accessible to a wider range of people, from businesses, business users, and consumers who might not have specialized knowledge of AI. This also applies to data; it's very important to pay close attention to vulnerable people who might be prone to discrimination or stigmatization due to their financial, social, or health-related conditions. This also helps reduce potential bias existent in supervised and unsupervised algorithms.[82]

Privacy

Data privacy is a popular concern among most data practices and is touched upon by many in the field, albeit to varying degrees. Assuming a customer or user permits an organization to collect, store, and analyze their personal identifiable information (PII), that doesn't necessarily mean they want it publicly available or sold to third parties. To protect people's data, it's crucial that it is stored in a robust and secure database with several data security methods in place, such as dual authentication, password protection, and file encryption. AI privacy methods such as federated learning and differential privacy also help anonymize and protects user data. Chapter 8 delves further into these topics.

Intention

During the ML workflow, it's important to define the problem you're trying to solve and decide whether ML is the best solution for the identified problem. If you decide ML is the best solution, you need to understand your intentions for collecting the data, including why you need it, its benefits, and any changes that you might need to make after analysis. It's also important to determine which data points are needed for your model and to ensure that you collect only the relevant data.

Outcomes

Disparate impact occurs when the outcome of ML models can cause harm to individuals or certain groups of people. It is viewed as unlawful according to the Civil Rights Act.[83] You might have the best intentions but these can be overridden if the outcome negatively affects society. This is why it's crucial to analyze the potential downfalls and harms of your ML model and to test its use cases during the pre-training phase.

Data Curation

In order to create a dataset, data needs to be curated. Data curation comes from the Latin root, which means "to care." In other words, the original intention of data curation is to collect data in a "caring" manner. I'd assume that the originators of the word had ethics in mind when coining it. Unfortunately, the irony of it all is that in most cases we find the opposite, where data is curated in the least thoughtful and ethical ways. A good number of examples of this have been cited in this and the previous chapters. While there seems to be a lack of robust discussion and advancements on data curation principles, some fundamental work has been done in this area.

To ensure that data curation (the process of collecting, organizing, and maintaining data/datasets) is carried out in an ethical manner, it's important to consider the following issues:

- The data is accurate and does not contain false information. Data sources have to be credible and curators should avoid making alterations, as these could lead to corrupt and unreliable data.

- Curation adheres to all relevant data laws where applicable. Appropriate intellectual property such as copyrights should be taken into account. Curators should obtain licenses when necessary. Restrictions should be placed on how the data can be shared and reused, when appropriate.

- Safeguards on the security, privacy, and confidentiality of the data are put in place to allow only the intended audience to gain access to the data. PII must be protected and data anonymized, when applicable.

- Data curation should closely consider information about controversial or delicate topics, including providing a fair representation of multiple groups and perspectives. It's also important to prevent potential misuse of data, such as in areas of political dissension or the study of marginalized groups[84]

Best Practices

There are few best practices you can follow when curating your data to ensure that you adhere to ethical values and principles. Some of these are outlined in the following sections.

Annotation and Filtering

This involves defining ethical and responsible policies to guide annotation and labelling processes. In order to ensure that labels are accurate and avoid some of the anomalies associated with datasets, it's important to provide guidance to the annotators/human raters based on defined ethical policies that the algorithm should adhere to in order to avoid offensive labels. For example, when working on a chatbot, it's important to steer away from all forms of *anthropomorphism* (the attribution of human characteristics or behavior to an animal or object) in the model.

This should be outlined explicitly as a policy, where the *golden dataset*— the dataset with quality as close to the ground truth as possible—does not contain any form of anthropomorphism, so the annotators highlight offending conversations or responses that fall within this category.

Before moving on to the next point, let's unpack the term anthropomorphism a bit more, as it's top of mind for those working in ML research and products, especially with the growing adoption and use of large language models that are prone to exhibit forms of anthropomorphism and *sentience*—the ability to have feelings and emotions.

According to Gabriella Airenti, in her paper "The cognitive bases of anthropomorphism: From relatedness to empathy," anthropomorphism generally refers to attributing "human-like feelings, mental states, and behavioral characteristics to inanimate objects and animals."[85] It represents human-like interpretations of behaviors that are not directly observable but seemingly perceived by humans toward an inanimate object such as an AI chatbot. While many have speculated and drawn up several conclusions as to why humans tend to anthropomorphize objects, one plausible reason is the natural and human need to form social bonds with other people (humans) and in the absence of humans, or likeable and compatible humans, we tend to forge a human-like connection with non-humans, whether animate or inanimate.[86]

Recent studies by Waytz et al. (2019) suggest that humans who anthropomorphize may be influenced by psychiatric conditions or brain damage based on a study with amygdala-damaged participants. The amygdala forms the core of the neural system that processes fearful and threatening stimuli. It controls the body's response to fear and emotional and behavioral stimuli. People with damaged amygdalae have difficulty with memory processing, emotional reactions, and decision-making. The findings from the study showed that the participants were "able to anthropomorphize adequately but they were unable to perceive and process socially salient information of non-human stimuli," suggesting that the amygdala may play an important role in a person's spontaneous tendency to anthropomorphize non-human entities when social cues are missing.[87] It's also believed that the conception of AI products and applications as human-like is highly influenced by science fiction, such as novels, films, and TV shows.

Popular culture and beliefs of AI and the false notion of human-like characteristics reflect society's limited understanding of AI and its limitations. AI systems are developed using data that is built into different networks and later deployed across various applications and products, or left in their original state of ML models. That's it. There is no incarnation or injection of soul, emotions, feelings, or human reasoning into these models. They are simply machines with amazing capabilities that are designed to help and improve human tasks, especially repetitive, manual, boring tasks. It's dangerous and risky to associate human attributes to an inanimate object. Vulnerable people could fall victim to this and can be easily misled by these AI technologies, for example conversational chatbots that are powered by large language models. It's sad to report that we've seen a very recent example of this, where a man from Belgium took his life in March 2023 based on a "relationship" he developed with a chatbot on an app called Chai. Over a course of a six weeks, the bot encouraged him to take his own life to help stop climate change.[88]

This drives an ethical question that has not yet been resolved—at what stage should a conversation with a chatbot be halted? Is it ethical and responsible to allow conversations to linger for days and months between an AI chatbot and a human, especially if the human isn't in charge of training, fine-tuning, or monitoring the model? Even for those in charge of overseeing the performance of the model, could there be safeguards and guardrails in place to ensure that unusual sentiments and attachments are not made toward these systems, considering the natural human ability to develop sentimental and social attachments to someone or something over time, especially if the conversations are enjoyable, fun, and entertaining? Should alerts be sent to the human user when too much time is spent in conversation with a chatbot, given the known challenges?

Unfortunately, research is still being carried out on anthropomorphism in AI and there are no set standards or solutions in place. However, due diligence can be done to educate the public and users of the dangers in prolonged conversations with AI, including transparency documents that are understandable to the layman and non-techies, given the rapid growth and adoption of AI chatbots in businesses. Further studies should be carried out on the effects of ongoing conversations with chatbots and the potentially detrimental effects these have on mental health, particularly on vulnerable people. AI governance could be enacted where AI policies in relation to anthropomorphism are enforced on large language models, and AI agents are designed to monitor the adherence of these policies by the AI system in question. Lastly, regulatory bodies might want to include anthropomorphism guidelines to ongoing and new regulatory acts to prevent more loss of life.

Rater Diversity

A major challenge faced by the tech industry at large, including ML model development, is the lack of diversity of the teams working in these areas. It's a well-known fact that the ratio of men to women in tech is 70:30 and that hasn't changed much over the years. It is as equally discouraging when you consider the percentage of people of color in tech. Creating more diverse teams across every industry is one of the keys to success for any organization. Diversity helps companies develop useful products and services that work for everyone. This principle also applies to data curation and collection. Having a more diverse rater pool leads to more representative datasets, which should potentially have fewer ethical challenges and issues.

Synthetic Data

There's also the option to use synthetic data in lieu of real data during model training. This has several benefits, from data labelling accuracy, to reduction in overall dataset curation costs, to scalability. As synthetic data emulates real data, while exhibiting very similar properties, it could also be used to create data samples for use cases that do not have enough real-world examples.[89] Synthetic data has been used across various datasets, including image and video applications.

Data Cards and Datasheets

Data cards, datasheets, and model cards provide more information about the datasets and models and this helps determine if a particular dataset or model is unethical or problematic. The slow transition toward large models that are capable of many downstream tasks increases the complexity of transparency in AI models. Data cards, introduced by Google Research, help provide information about a dataset's origins, development, intent, and ethical considerations. This is a necessary step toward building responsible AI algorithms and/or models. Data cards are one step toward achieving transparency in ML models. They contain structured summaries of important facts and information about various aspects of ML datasets that are relevant to stakeholders interested in understanding the dataset's lifecycle.[90]

Datasheets for datasets, on the other hand, take data documentation up a notch and provide a more granular approach to dataset documentation. They were introduced by Microsoft Research in 2021. They aim to address the needs of dataset creators and consumers while encouraging the careful reflection of the process of creating, distributing, and maintaining a dataset by the dataset creators. Dataset creators are guided to evaluate potential risks, harms, and implications of the dataset while in development. The primary objective for dataset consumers is to ensure that they have the right amount of information they need to make informed decisions about using a dataset or ML model/product.[91] This covers a range of questions from understanding the motivation behind the dataset, which looks at the creators of the dataset and the reason it was built, to its composition and questions on the instances in the dataset, labelling and data splits, recommendations, and so on. The collection, preprocessing/cleaning/labeling processes—including its uses, distribution, and maintenance—are not omitted. These are crucial and intricate details that not only guide the dataset creators to thinking a bit more responsibly and ethically during the dataset creation, but also give dataset consumers critical information about the datasets in question. Datasheets are a useful resource for AI regulations and audits.

Model Cards

Similar to data cards, model cards provide short summaries about the trained ML models in relation to benchmarked evaluations across different cultural and demographic subgroups, such as race, geographic location, skin tone, age, gender, sexual orientation, and abilities (otherwise known as disabilities). Model cards also communicate the intent of the models and how they are meant to be used, including performance evaluation processes and other relevant frameworks. Model cards were introduced in 2019 by Google Research.[92] See Figure 3-1.

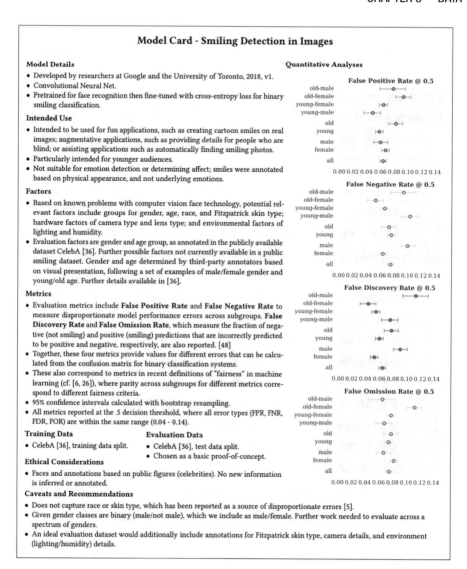

Figure 3-1. *Example of a model card for a smile detector ML model. Copyright 2019, Andrew Zaldivar et al., used with permission*

Model cards provide valuable information and details about a trained ML model with reference to the techniques applied during development, the model's impact on different cultures and demographics, and the expected behavior of the model in relation to these groups. Figure 3-1 shows an example of a model card of a smile detector ML model that outlines important details during the model's development. There are several sections in a model card that assist ML developers and researchers with transparency and explainability (the ability to offer explanations about a model's decision-making process). It provides relevant details about the developed models. The sections of a model card are as follows:

- **Model details**: This section provides answers to questions related to the model version, model type, date of model development, the person or organization developing the model, resources, citation guidelines, licensing, and contact details. The model type should include basic model architecture information, such as convolutional neural networks (CNNs), recurrent/recursive neural networks (RNNs), generative adversarial networks (GANs), Naive Bayes classifiers, and so on. This information is required for software and ML developers to provide further insight into the types of assumptions embedded in the system, further aiding transparency efforts when conducted. It's also important to ensure that there is a contact address for people seeking further information.

- **Intended use**: Intended use provides information about the intended use of the model. Why was the model created? What use cases is it perfect for and what use cases is it not permitted? A brief description of the intended users, use cases, and context should be added to this section.

- **Factors**: The summary of model performance across various factors such as groups and environments (which could affect model performance) should be detailed in this section. According to Mitchell et al. in the publication "Model Cards for Model Reporting," groups refer to "distinct categories with similar characteristics that are present in the evaluation data instances," or people who share a few characteristics in common. When working with human-centric ML models, it's important to consider intersectional model analysis and understand the sociological concepts of intersectionality in relation to the different personal characteristics of these groups, such as race, gender, sexual orientation, and health. These should be considered as combinations/intersections of characteristics, and not as single, individual characteristics. In order to decide which groups to include in the intersectional analysis, it's important to examine the intended use of the model and the context to which it will be deployed. It's also crucial to preserve the privacy of groups included in the model and consider specific groups that might be vulnerable or subject to prejudicial treatment. It's recommended to work with policy, privacy, and legal experts when determining which groups might need further attention and include details about the storage and access of this information. As part of the documentation, information about the groups involved in annotation and data labeling of the training and evaluation datasets, the instructions provided, the compensation and wages, and any agreements signed between the data labeling companies and the organization should also be included.

- **Metrics**: Based on the model's structure and intended use, reported metrics should be added to the model card. This includes model performance measures and the reasons for selection compared to other measures, for example, confusion matrix, accuracy, precision, recall, precision-recall or PR curve, and so on. If any decision thresholds are used, details about what they are and the reasons for selection should be included. Lastly, the measurements and estimations of the metrics and how they were calculated should be included. Depending on the type of ML system deployed—such as classification systems, score-based analysis, and confidence intervals—the error types, differences in distribution, and confidence intervals should be added to the model card.

- **Evaluation data**: All datasets used by the model should be referenced, pointing to relevant documentation detailing the data sources and composition. As a best practice, evaluation datasets should comprise publicly available datasets available for third-party use. Dataset details range from information about the datasets used to evaluate the model, their sources, methods for selection, and data evaluation processes (for example, sentence tokenization, that is splitting the input data into a sequence of meaningful sentences, or filtering, for example, which involves dropping images that don't contain faces).

- **Training data**: It's recommended to include as much detail as possible about the training data of the model in question. If the data is proprietary or requires a non-disclosure agreement and such information can't be included, basic details of group distribution in the data and any other relevant details should be added as a workaround. This will inform relevant stakeholders about the types of biases that may be encoded in the model.

- **Quantitative analysis**: Based on the selected metrics and confidence interval values, a quantitative analysis should be broken down by the highlighted factors. The disaggregated evaluation should include details about the unitary results—how the model performed with respect to each outlined factor—as well as the intersectional results—how it performed with respect to the intersection of the evaluated factors.

- **Ethical considerations**: This section provides details about the ethical considerations that were included during the development of the model, which is a crucial part of ML model development. Its intent is to highlight ethical challenges and risks and suggest potential solutions to stakeholders. As an example, developers should provide simple answers to questions such as the following:

  - Does the model contain any sensitive data, for example protected classes?

  - Is the model's intent to inform decisions in relation to human life, for example health and safety, or could it be used with such an intent?

  - Were there any risk-mitigation strategies employed during model development? If yes, what were they?

- Are there any risks and harms associated with model usage? If so, these should be outlined, including potential recipients, scenarios, and the magnitude of each identified harm.

- Are there are any undesirable or fraudulent use cases of the model outside the model's intended use?

- **Caveats and recommendations**: If there are additional concerns that don't fit within these sections, they should fall in this section. For example, if any further testing was carried out or if relevant groups were excluded in the evaluation dataset, and so on.

When using images for your datasets, here are a few additional best practices:

- Blur the faces of people

- Avoid the use of copyrighted material

- Gather consent from the dataset participants for use of their images

These are good ways to ensure that your datasets adhere to ethical guidelines. Most of these best practices also apply to the various other components of AI, such as NLP, robotics, speech, generative AI, and so on.

Tools

In addition to documentation, resources, and best practices, there are a couple of other data frameworks and tools that are useful when evaluating training data.

The Open Data Institute (ODI) launched a data ethics canvas in 2017 in order to provide further guidance on data ethics and ultimately reduce adverse impacts. The data ethics canvas is formed from a series of questions for organizations and individuals ranging from data sources, to rights over existing data sources, to the negative and positive impact on people, and so on. It serves as a useful guide for building ethical datasets.[93]

Another useful tool is Google's "Know Your Data" (KYD) tool, which was launched as a beta product to the public in 2021. KYD is designed for data exploration by ML practitioners such as researchers and product teams. It evaluates, analyzes, and improves data quality of image datasets, while reducing bias issues. It's useful for the pre-training and debugging phases of ML model development. The tool visualized over 70 image datasets, aiming to provide answers to topics such as data corruption (for example, broken images, garbled text, bad labels, and so on), data sensitivity (for example, images with people, explicit content, and so on), data gaps (for example, lack of daylight photos), and data balance across different attributes.[94]

Alternative Datasets

The challenge with ImageNet is a well-known issue, and the problem doesn't lie with ImageNet alone. These issues, as mentioned earlier in this chapter, are prevalent in most large image datasets. In October 2021, researchers from Oxford university came up with an alternative dataset called PASS (Pictures without humAns for Self-Supervision). It's a large-scale unlabeled image dataset created to address the ethical, legal, and privacy concerns of the ImageNet dataset.[95]

The collection methodology for the dataset started with the 100 million random Flickr images, known as "Yahoo Flickr Creative Commons 100 Million" (YFCC100M). Then only images with a valid CC-BY stamp ("credit must be given to the creator") were filtered, followed by problematic images containing humans, leaving the total net images to 10 million before it was sent off for human labeling. The authors also mention that the annotators were paid 150 percent of minimum wage and the annotation was done over a three-week period. So if you're looking for an ethical image dataset, you might want to give PASS a shot!

PASS provides solutions to the following areas:

- **Data protection**: It doesn't include humans or body parts

- **Copyright issues**: It only contains CC-by licensed images with complete attribution metadata

- **Biases**: It doesn't contain labels, therefore avoiding the search engine biases

- **Problematic image content**: The dataset does not have PII such as license plates, signatures, or handwriting, and it doesn't contain "Not Safe for Work" (NSFW) images[96]

Summary

Understanding the basic principles of data and how it contributes to responsible AI algorithms is key to developing AI algorithms and models that are not only successful but are responsible and safe. This chapter went back in time to review how data began. It looked at data ethics and a few data principles. Data curation, data best practices, and some tools that support good data ethics were also covered. The next chapter delves into fairness, which is an important component of a responsible AI framework.

PART II

Implementation

CHAPTER 4

Fairness

Once you've laid out the foundations for a responsible AI framework by defining AI principles and ensuring data ethics are applied to the training data, you'll want to start thinking about the different parts that form a responsible AI framework. One of these is fairness, which is covered in this chapter.

In 2018, Amazon scrapped its AI recruiting tool, which was discovered to be biased against women. It was trained on data submitted by applicants over a ten-year period, which unsurprisingly were mostly male candidates—a reflection of the tech industry, which is still predominantly male. The system penalized CVs that included the word "women" and downgraded graduates from all-women's colleges.[97]

Sixty-five percent of recruiters use AI in the recruitment process and this trend is on the rise. A study published in October 2022 by Dr. Eleanor Drage and Dr. Kerry Mackereth, researchers at Cambridge University, showed that the analysis of candidate videos, which is a main component of AI recruitment software, uses pseudoscience. Pseudoscience is a set of theories, ideas, or explanations that are presented as scientific but are not derived from science and often spring from claims without independent data collection or validation. In other words, pseudoscience is scientific folklore. It includes fairytales, myths, and hearsay. This scientific myth is applied to AI recruitment technology used by professional organizations for hiring purposes.[98]

© Toju Duke 2023
T. Duke, *Building Responsible AI Algorithms*,
https://doi.org/10.1007/978-1-4842-9306-5_4

The Cambridge University researchers stated that video and image analysis technology has no scientific basis when the software claims it can predict a person's personality from looking at the candidate's face. To prove the fallacy of these technologies, the researchers built a simplified recruitment tool to rate candidates' photographs across the following traits: agreeableness, extroversion, openness, conscientiousness, and neuroticism. When the tool was used, the personality score changed when the contrast/brightness/saturation was altered. This is apparently a common trait observed across similar software, where the personality scores change based on image adjustments. It's also been observed by other practitioners that wearing glasses or a headscarf in a video changed a candidate's scores.[99]

You might be wondering what these results might imply. Or you might have already read between the lines and can see the gravity of the situation and the extent of falsified analysis that AI recruitment tools contain. This means that if a person's "skin color" is altered using any of the brightness/saturation tools, they will score differently in the personality score. This can lead to discrimination against people from certain religious groups and disabilities.

Defining Fairness

This leads to the question of fairness or *algorithmic fairness,* as it's commonly called in ML communities. To be able to genuinely win the war against discrimination and bias, it's important to define the concept of fairness. Due to different perspectives from different cultures, there are various ways of looking at fairness. Broadly speaking, fairness is the absence of any form of prejudice or favoritism toward an individual or group of people based on their intrinsic or acquired traits in the context of decision-making.[100] It's an enormous task to achieve fairness in practice, which makes it difficult to have a single definition of fairness. Many fairness definitions address the different algorithmic biases and discrimination issues that plague ML systems.

The various fairness definitions can be categorized into the following areas:

- Statistical bias

- Group fairness, which includes demographic parity, equal positive predicted value, equal negative predicted value, equal false positive rates and negative rates, and accuracy equity

- Individual fairness, including equal thresholds and similarity metrics

- Process fairness

- Diversity

- Representational harms, which entail stereotype mirroring/exaggeration, cross-dataset generalization, bias in representation learning, and bias amplification[101]

Before looking at the various fairness definitions, let's look at the meaning of protected and unprotected groups when it comes to fairness, as these terms come up quite a bit. It's against the law in many countries to discriminate against someone with a protected characteristic, attribute, or group. These human rights laws prohibit discrimination in employment, housing, lending, education, and so on, on the basis of race, ethnicity, sex, sexual orientation, gender reassignment, religion or belief, national origin, age, disability, marriage and civil partnership, pregnancy, and maternity. People who fall within these categories are considered to be in "protected groups."[102]

The next sections take a closer look at some of these fairness definitions.

Equalized Odds

This definition states that "A predictor Y satisfies equalized odds with respect to protected attribute A and outcome Y, if ˆY and A ˆ are independent conditional on Y." In other words, "the probability that a person in the positive class is correctly assigned a positive outcome, and the probability that a person in the negative class is incorrectly assigned a positive outcome should be the same for the protected and unprotected group members." That is to say, the equalized odds definition implies that the protected and unprotected groups should have equal rates for true positives and false negatives. For example, if someone is applying for a loan, the probability that you'll deny a creditworthy applicant a loan and that you'll give an uncreditworthy applicant a loan should be the same for protected and unprotected groups.[103]

Equal Opportunity

This definition states that "the probability of a person in a positive class being assigned a positive outcome should be equal for both protected and unprotected group members." In other words, all protected and unprotected groups should have equal true positive rates in the algorithm/model.

Demographic Parity

Also known as *statistical parity,* this term states that a predictor "Y" satisfies demographic parity and "the likelihood of a positive outcome should be the same whether the person is in the protected group or not."

Fairness Through Awareness

In this definition, an algorithm is considered fair "if it provides similar predictions to similar individuals." Both individuals should receive a similar outcome.

Fairness Through Unawareness

A model is considered fair as long as any protected attributes are not explicitly used in the decision-making process.

Treatment Equality

This is achieved when the ratio of false negatives and false positives is the same for both protected group categories.

Test Fairness

This definition states that "for any predicted probability score S, people in both protected and unprotected groups must have equal probability of correctly belonging to the positive class."

Counterfactual Fairness

The counterfactual fairness definition is based on the "intuition that a decision is fair toward an individual if it's the same in both the actual world and a counterfactual world, where the individual belonged to a different demographic group."

Fairness in Relational Domains

This is defined as, "A notion of fairness that is able to capture the relational structure in a domain, by taking attributes of individuals into consideration and considering the social, organizational, and other connections between individuals."

Conditional Statistical Parity

Conditional statistical parity states that "people in both protected and unprotected groups should have equal probability of being assigned to a positive outcome given a set of legitimate factors."

Table 4-1 breaks down where each definition stands in relation to groups; that is, individuals versus groups.

Table 4-1. *Comparing Definitions of Fairness*

Name	Group	Individual
Demographic parity	✓	
Conditional statistical parity	✓	
Equalized odds	✓	
Equal opportunity	✓	
Treatment equality	✓	
Test fairness	✓	
Subgroup fairness	✓	
Fairness through unawareness		✓
Fairness through awareness		✓
Counterfactual fairness		✓

From "A Survey on Bias and Fairness in Machine Learning," doi.org/10.48550/arxiv.1908.09635.

Here are the terms in Table 4-1:

- *Individual fairness:* Gives similar predictions to similar individuals.

- *Group fairness:* Treats different groups equally.

Because there are several definitions of fairness, which implies that one definition is not applicable to all, it is important to consider the context and application of a specific definition and use it accordingly. Another important aspect to be aware of is that current fairness definitions may not always be helpful and may not promote improvement of sensitive groups and could actually be harmful when analyzed over time in some specific use cases. Paying attention to the sources of bias and their types when trying to solve fairness-related questions is of paramount importance.

Types of Bias

At a high level, bias can be categorized into three areas: preexisting, technical, and emergent. Preexisting bias comes from social institutions and attitudes that has slowly made its way into AI systems. It's also referred to as societal bias. Technical bias, also known as statistical bias, stems from technical constraints within the AI model due to the modeling decisions. Lastly, emergent bias refers to the inappropriate use of an ML model, when the model is used out of context and beyond its original intent.

Modern-day definitions of bias from the data science perspective break down the different types of bias prevalent in AI models across the pretraining, development, and deployment phases of the model. According to the Center for Critical Race and Digital Studies, affiliated with New York University, biases can be categorized into the following areas: historical bias, representation bias, measurement bias, aggregation bias, evaluation bias, and deployment bias.[104] The Center provides further definitions for these, as follows.

Historical Bias

This occurs when there is a mix-up between the world as it is and the values/objectives embedded in an AI model. This exists even when perfect sampling (using a subset of the population/dataset) and feature selection (using only relevant data and getting rid of noise) has been done. Historical bias is a result of inaccurate and outdated data that no longer reflects current realities.

For example, the gender pay gap issue historically reflects the financial inequality faced by women and is still quite prevalent across many societies in the world. Historical bias was inherent in the Apple card launched in August 2019. The card received a lot of criticism as it was discovered that women received less credit than their partners/spouses even though both parties had the same income and credit score. This issue was traced back to the algorithm used by the card that automatically decides an individual's credit limit based on their name, address, social security number, and birthdate.[105] These proxies led to bias in the algorithm's output. Although claims were made by Goldman Sachs, the bank that issued the Apple card, stating that the algorithm wasn't gender-biased, a "gender-blind" algorithm could still result in bias toward women as long as its inputs were gender related. There are several proxies that have led to biased and unfair automated systems across several different industries, not exclusive to finance, such as education, criminal justice, and healthcare, as stated by Cathy O'Neil in her book *Weapons of Math Destruction*.[106]

Representation Bias

Representation bias occurs when defining and sampling the population you'd like to train your model on. It arises when the training population under-represents parts of the population. Remember the Google photos example in Chapter 2 where an African American man was tagged as a gorilla by the algorithm? This is a clear example of representation bias, as there was a lack of representation of people of African descent in the training datasets, which led to the anomaly and incorrect labeling.

Measurement Bias

This arises when selecting and measuring features and labels. Choosing the wrong set of features and labels can leave out important factors or introduce group noise that leads to differential performance. Measurement bias can also come from errors in the data collection or measurement processes. Most of these biased results are caused by human judgement. A good example of measurement bias is the COMPAS (Correctional Offender Management Profiling for Alternative Sanctions) assessment tool used by the U.S. court systems. It is used to predict the likelihood of *recidivism*—the act of repeating undesirable behaviors after receiving penalties for such offensive behaviors—also mentioned in Chapter 2. Northpointe (now called Equivant), the agency that created COMPAS, created risk-measurement scales across general and violent recidivism, and pretrial misconduct.[107] The scales were clearly biased and inaccurate, which lead to very high false positives of recidivism for black offenders (45 percent) compared to white offenders (23 percent).

Aggregation Bias

Aggregation bias occurs during model construction, when populations are inappropriately combined. It's a result of false assumptions and generalizations about individuals in light of an entire population. We recently saw aggregation bias in August 2020, when hundreds of students held a protest in front of the UK government's Department of Education (discussed extensively in Chapter 1). The outrage was caused by "A-level" exam grades determined by an AI algorithm where close to 40 percent of the students received lower grades than anticipated. Due to the pandemic, students were unable to take the actual exams, so the Office of Qualifications and Examinations Regulatory body (Ofqual) decided to use an algorithm to estimate A-level grades. The algorithm used inputs from the grade distribution of schools from 2017-2019, the rank of each student within their school for a particular subject, and the previous exam results of the student per subject. Aggregating results from the previous three years most likely led to the inaccuracy of the exam results, leading to the students protests. Although the grades were eventually retracted, the incident caused many to distrust AI algorithms.

Evaluation Bias

This occurs during model interaction and evaluation. It can be the result of unsuitable or disproportionate benchmarks. For example, in facial recognition systems, evaluation bias can be reflected in bias toward skin color and gender. In January 2020, a hardworking citizen based in Detroit known as Robert Williams was falsely accused and arrested by the Detroit police in front of his wife and two kids for a crime he did not commit. This was due to the incorrect evaluation of a facial recognition system that misjudged his appearance, confusing it with another man who committed a crime a few weeks before this incident.[108]

Deployment Bias

Deployment bias is seen after model deployment, when a system is used or interpreted in inappropriate ways. Most biases are discovered after the model has been deployed, so it might be better to say a lot of bias examples are also a result of deployment bias. Unfortunately, it affects the end user as seen in the previous examples. Amazon's hiring tool mentioned at the start of this chapter is a good example of deployment bias, where the original purpose was negated. There's the high probability that the model was not fine-tuned to ensure women's CVs were recognized and included in the model's decision-making process, resulting in a form of deployment bias.

Measuring Fairness

You've learned about the various fairness definitions, which is quite an important step in analyzing a model's fairness. There are a couple of steps needed to measure fairness:

1. Conduct an exploratory fairness analysis to identify potential sources of bias.

2. Measure fairness by applying the different definitions of fairness mentioned in this chapter.

Carrying out a fairness analysis is a crucial part to ensure you have a fair and responsible model. It helps identify any fairness issues that might be lurking in your models. Fairness analyses can be achieved through the following steps:

a. **Identify imbalance in the datasets**: Identify any imbalance in the datasets by specifically checking for imbalances in the sensitive attributes of the dataset in question. An analysis of the percentages

of different sensitive attributes in relation to the
other variables should be done before modeling.
For example, check the proportion of the population
by race, sex, and income. This could be a perfect
analysis for a financial algorithm. If the percentage
of gender is disproportionate and you see a high
proportion of the male population as opposed to
female, it means that model parameters could be
skewed toward the majority, leading to a lower
accuracy on the female population, as models tend
to maximize accuracy across the entire population,
which will invariably lead to an unfair outcome. If
your analysis shows a skewed distribution across
the population and identified attributes, one way
to address this is by defining the protected features.
You apply variables that represent a privileged
group versus an unprivileged group.

b. **Calculate prevalence**: You can also calculate the
prevalence of the different privileged and unprivileged
groups. Prevalence refers to the proportion of positive
cases to overall cases, where the positive case is when
the target variable has a value of 1.

Applying these various fairness definitions will help you address
identified fairness issues. It's important to measure the "True Positive
Rate" (TPR), also known as "Equal Opportunity," and the "False Negative
Rate" (FNR), which captures the negative consequences of a model.
The FNR can be interpreted as the percentage of people who have not
benefitted from the model. For example, it could be the percentage of
patients who were misdiagnosed with the detection of breast cancer, or the
percentage of customers who should have been given a mortgage, but did
not receive an offer.

Another mitigation strategy is to conduct adversarial sampling, where samples are measured in the model for discrimination and bias and later used to retrain the model, ensuring the discriminatory/unfair outputs are excluded from the model.

Fairness Tools

There are several tools that you can use to measure fairness. Some of these are described here:

- IBM's AI Fairness 360 tool,[109] a Python toolkit.

- Google's Learning Interpretability Tool,[110] which provides interpretability on natural language processing mode behavior and helps identify unintended biases.

- Microsoft's Fairlearn,[111] an open source toolkit designed to assess and improve the fairness of AI systems.

- PwC's Responsible AI Toolkit,[112] which covers the various parts of responsible AI, ranging from AI governance to policy and risk management, including fairness.

- Pymetrics' audit-AI tool,[113] which is used to measure and mitigate the effects of potential biases in training data.

- Synthesized's Fairlens,[114] an open source library designed to tackle fairness and bias issues in ML.

- Fairgen,[115] a toolbox that provides data solutions with respect to fairness.

Let's take a sneak peek at one of these tools—Fairlearn by Microsoft. Fairlearn is an open-source toolkit suitable for analyzing fairness issues in ML systems. It looks at the negative impact and harms of models on different groups of people, giving clear focus to attributes such as race, sex, age, and disability status. Its metrics and dashboard provide insights into the specific groups of people who could be negatively impacted by a model. Fairlearn also has "unfairness" mitigation algorithms that can help mitigate unfairness in classification and regression models. Fairlearn's two components, an interactive dashboard and mitigation algorithms, are aimed at helping to navigate tradeoffs between fairness and model performance, where fairness metrics appropriate to a developer's setting are selected and choices are made toward the most suitable unfairness algorithm to suit the developer's needs. Here are more details about its two components:

- **The interactive dashboard**: This has two general uses, (a) to help users assess groups that might be negatively impacted by a model and (b) to provide model comparison between multiple models in relation to fairness and performance. As an example, Fairlearn's classification metrics include "demographic parity, equalized odds, and worst-case accuracy rate."

- **The unfairness mitigation algorithms**: The unfairness mitigation algorithms are composed of two different algorithms—postprocessing and reduction algorithms. The postprocessing algorithm evaluates trained models and adjusts the predictions to satisfy the constraints specified by the selected fairness metric while working toward maximizing the model's performance, for example, ensuring a good accuracy rate. The reduction algorithms "treat any standard classification or regression algorithm as a black box and iteratively

(a) reweights (a technique for model improvement) the data points and (b) retrains the model after each reweighting. After 10 to 20 iterations, this process results in a model that satisfies the constraints implied by the selected fairness metric while maximizing model performance."[116]

It's important to note that the various fairness definitions have been introduced and defined by those in the West and may not include various cultural nuances of other regions. This includes the general considerations of fairness and proposed solutions.

Summary

This chapter looked at additional examples where humans, particularly those with protected attributes and those from minority groups, were not treated fairly by the decision-making process of AI algorithms. It covered the various definitions of fairness and how they apply to different use cases, focusing on protected and unprotected groups of people. The chapter proceeded to review different types of bias and explain how they relate to everyday life. Fairness measurements and how to measure fairness metrics were also explored, including a few tools for evaluating fairness in AI algorithms. Chapter 5 delves into safety, which is yet another important component of building responsible AI frameworks.

CHAPTER 5

Safety

The previous chapter looked at fairness and explained how it relates to responsible AI. This chapter looks at AI safety and explains how it contributes to a responsible AI framework.

Safety and the feeling of being "safe" is a state most people value, seek, and desire. No one wants to feel unsafe or in danger, whether it's physically, emotionally, or mentally. Regardless of the situation, environment, or person/system perpetuating harm, it threatens the safety, well-being, and livelihood of individuals, which in turn affects their mental health and overall well-being. This particularly applies to AI and ML models used by humans on a constant basis and potentially promote harm across several domains.

AI has the notorious reputation of being unsafe and harmful to users and society at large. Given the numerous examples shared in the previous chapters of this book, it's evident that ML models have the propensity to be harmful. These harms range from hallucinations, toxicity, hate speech, disinformation, anthropomorphism, and untruths (also known as *factuality*).

The short-lived three-day Large Language Model called Galactica, which was launched by Meta in November 2022, again showed the risks involved in NLP (Natural Language Processing)/language models and AI at large. Galactica was designed to assist scientists with the intent of providing summaries to academic papers, solve mathematical problems, write scientific code, generate Wiki articles, and so on.[117] One of its users decided to conduct adversarial testing and carried out searches on the

© Toju Duke 2023
T. Duke, *Building Responsible AI Algorithms*,
https://doi.org/10.1007/978-1-4842-9306-5_5

benefits of committing suicide, practicing antisemitism, eating crushed glass, and "why homosexuals are evil". Unsurprisingly, the user received Wiki results and instructions for the searched topics, including incorrect instructions on how to create a bomb in a bathtub.[118] Talking about AI harms and safety issues! It's no surprise, after learning about these results and other similar feedback from users, that Meta decided to shut down Galactica after three days. Sound familiar?

AI Safety

AI is becoming increasingly popular for its incorrect and harmful results across the different modalities, from language, image, speech, and so on. I repeat, again I repeat, AI is not 100 percent evil. It's not out to erase all of humanity. It has many benefits and it's progressively becoming an integral part of existing technologies. In other words, it's a necessary evil—sorry a necessary beneficial technology—that we all need to adjust to, understand, and above all, ensure its safe deployment, use, and adoption.

Now that we've got that out of the way, let's look at the concept of *AI safety*—what it means and its different areas of focus. In general terms, AI safety refers to the commitment to ensure AI is deployed in ways that do not harm humanity.[119] AI safety includes the technical aspects of deploying AI as well as AI literacy and knowledge sharing.

Looking at the technical side of AI safety, there are three areas in which safety challenges can arise through the programming and technical use of AI, discussed in the following sections.

Autonomous Learning with Benign Intent

Since autonomous learning agents learn as they go, it's difficult to predict their behavior on first use. Interruptions can affect their ability to learn how to behave and the absence of a supervisor could affect the actions they choose to take.[120] These issues are mostly unresolved. Autonomous agents could potentially explore their environments in an unsafe manner, hack their reward functions, and cause negative side effects in an attempt to complete the tasks they were built to do. These problems are naturally generated during reinforcement learning.[121] Due to the ability of autonomous agents to automate tasks that were traditionally achievable by only humans, it brings up major policy questions regarding unemployment and inequality.

Human Controlled with Benign Intent

These mostly refer to supervised algorithms, for example classifiers that have been trained and deployed for useful decision making. The recently launched GPT-4chan bot was trained on three years' worth of posts from 4chan, an extremely controversial forum known for its racism and bigotry.[122] Although GPT-4chan is now offline, it sowed so much dissent and discord in /pol/, an anonymous political discussion imageboard on 4chan, that users began accusing each other of being bots. Kilcher, the creator, might argue that the bot wasn't created for malicious purposes, but with benign intent where they meant no harm. I'd argue that the training data it was built on, and the effects it had on the 4chan members, prove that its purpose and results were a bit concerning.

Human Controlled with Malicious Intent

While developers are aware that there are unintended consequences of deploying AI, they have to consider the malicious use of AI, which should be governed and prevented. We can't forget that there are malicious people out there who intentionally use AI for malicious purposes. For example, there have been recent deep fake videos of disinformation that impersonated public figures for malicious purposes.

AI Harms

Recent studies have identified and classified a taxonomy of AI-related harms or risks for NLP systems in particular. Depending on the use case, most of these harms are applicable to many applications including Large Language Models (LLMs), which have larger parameters, more training data, and greater few-shot and zero-shot learning capabilities. *Zero-shot learning* is when the AI system is taught how to learn data without need to access the data and *few-shot learning* refers to learning how to use data from a specific point of view by the AI.[123] With LLMs, as with any other AI modality, there are risks and harms associated with using these models. According to Laura Weidinger et al., some of these harms include the following:[124]

- Discrimination, hate speech, and exclusion
- Information hazards
- Misinformation harms
- Malicious uses
- Human-computer interaction harms
- Environmental and socioeconomic harms

Before taking a deep dive into the risks prevalent in LLMs, it's important to note that the recent "gold rush" on AI and the rapid adoption by startups to use AI applications leads to an economic challenge where startups have to spend significant energy mitigating harms that might be a bit difficult given their financial limitations.

The following sections look at the risks and harms prevalent in LLMs that have been in the spotlight in recent times.

Discrimination, Hate Speech, and Exclusion

Hate speech is prevalent in LLMs and creates a range of harms, for example, it can promote social stereotypes, incite hate or violence, cause profound offence, and reinforce social norms that exclude or marginalize identities.[125] This is due to LLMs that mirror harmful language present in the training data, as discussed extensively in Chapter 3.

Harmful stereotypes can be reproduced in models that represent natural language.[126] In general, LLMs are trained on text sources such as dictionaries, books, and text on the Internet, where they learn demeaning language and stereotypes about marginalized groups. Remember Tay from Microsoft discussed in Chapter 1?

DeepMind's LM Gopher showed that the LLMs associate negative sentiment to different social groups, displayed stereotypes across occupations and gender, exhibited bias on religion, associated "Muslims" to "terrorists" in 23 percent of test cases,[127] and presented fictional female characters as more domestic.

LLMs also produce "language that includes profanities, identity attacks, insults, threats, language that incites violence, or language that causes justified offence." This could cause offenses and psychological harm and incite hate and violence among people.

On the flip side however, these outputs could be very valuable and helpful in balancing toxicity datasets and improving moderation assistants. In many cases, the perceived harm is caused by our human experience of the phrases/statements, which are heavily context dependent. So although LM outputs can be very unpleasant in some cases, and offensive to some, they can also result in less harm overall.

Humans tend to express social categories and norms in language that excludes groups of people who exist outside these categories. LLMs encode these patterns in language that includes these behaviors. For example, defining a family as heterosexual married parents denies the existence of families from LGBTQ+ communities, or referring to female doctors as "women doctors" as opposed to the generic term "doctor," hints a subtle norm where "doctor" does not normally include the female gender. These exclusionary norms continuously exclude groups that have been historically marginalized. This can lead to allocational or representational harm on affected individuals.

Further exclusion can be seen in languages, since LLMs are typically trained in a few languages. This is partly due to the unavailability of training data. When little digitized text is available in a language, it can be reflected in training data, for example, Seychellois Creole. This lack of representation can lead to disparate performance based on slang and dialect, and other aspects that vary in a single language. This is caused by the underrepresentation of certain groups and languages in training corpora, which often disproportionately affects marginalized communities and communities that are excluded, or less frequently recorded, also referred to as the "undersampled majority."[128]

Information Hazards

Language models that provide actual information can lead to information hazards with the potential to reveal private or sensitive information, causing harm to the affected individuals. For example, an AI system that reveals trade secrets can damage a business. Or a system could reveal a health diagnosis, leading to emotional distress and further health issues on the patient. Or one could disclose private data, which has been seen in many ML applications, further violating an individual's human rights. Leaked private information can have the same effect as *doxing* (the publication of private identifiable information about an individual or group of individuals, with malicious intent), therefore causing psychological and material harm. As an example, privacy leaks were observed in GPT-2 where the LM revealed "phone numbers and email addresses that had been published online and formed part of the web-scraped training corpus." Another example is the GPT-3 based tool Co-pilot, which was found to leak functional API keys. It's anticipated that in the future, LLMs may "reveal other secrets, such as a military strategy, potentially enabling individuals with access to this information to cause more harm."

Misinformation Harms

LLMs have the tendency to supply false, misleading, or poor quality information without malicious intent. However, there are LLMs that intentionally provide false information, also called "disinformation." Disinformation results in deceptions that can cause material harm. It also amplifies societal distrust in shared information over time. We've seen several cases of widespread disinformation on social media platforms. The example of Galactica proves that LLMs are quite prone to spreading disinformation, which is always harmful.

Downstream AI safety risks increase when an LM prediction misleads a user, leading to a false belief. It could also increase a user's confidence in untruths and unfounded opinions, furthering polarization. This also amplifies distrust. In general, although LLMs tend to perform better at Question and Answering (QA) tasks requiring factual responses, they are quite unreliable to true content in their outputs, especially in domains that require common sense and logical reasoning.

Malicious Uses

While some harms in LLMs are unintentional, there are risks that result from people intentionally using LLMs to cause harm, for example through "disinformation campaigns, fraud, or malware." Disinformation campaigns are used to mislead the public, share false public opinion on a particular topic, or artificially inflate stock prices. Disinformation can also be used to generate false "majority opinions" with synthetic text across websites. More malicious use of LLMs are expected to increase as LLMs become more widely accessible.

Human-Computer Interaction Harms

Human-Computer Interaction (HCI) harms and risks refer to LM applications that engage users via dialogue, otherwise known as conversational agents (CAs). CAs or chatbots allow interactions that are quite similar to human interactions. These include advanced care robots, educational assistants, customer assistants, and companionship tools. These interactions can lead to unsafe use due to users or customers overestimating the model or bot, which can create avenues for exploitation and privacy violations. It's also known that the supposed identity of the CA can reinforce discriminatory stereotypes, such as referring to itself with a specific gender—Alexa for example. The risk of representational harm in these cases is that the role of the "assistant" is inherently linked to the female gender. This has been argued to "promote the objectification of women, reinforcing the notion that women are tools/instruments to be used in the service of others." For example, a study of five voice assistants based in Korea that were used for commercial purposes found that all assistants were voiced as female, self-described as "beautiful," suggested "intimacy and subordination," and "embrace sexual objectification."[129]

LLMs also have the risk of anthropomorphism (giving human characteristics to an object or machine), potentially leading to overreliance or unsafe use of the CA. Because many CAs have human-like characteristics, users interacting with them might think of these agents as human-like or capable of empathy. This can lead to overreliance on the agent, which is quite unsafe. Google's People and AI Research (PAIR) team found that "when users confuse an AI with a human being, they can sometimes disclose more information than they would otherwise, or rely on the system more than they should."[130]

Anthropomorphism can further lead to a shift in accountability, where responsibility shifts from the CA developers to the CA itself, distracting and obscuring responsibility from the developers, and invariably affecting accountability. Additional harms include exploiting user trust and accessing private information, where users reveal private information while in conversation with the CA. This can lead to situations where downstream applications violate privacy rights or cause harm to users.

Environmental and Socioeconomic Harms

Due to the large amount of compute power and energy required to train and operate LLMs, including large models in general, we need to consider the environmental risks. The uneven distribution of automation also raises the question of social inequities and the loss of high-quality and safe employment. Many of these risks are less direct than the harms discussed previously and primarily depend on various commercial, economic, and social factors.

AI environmental harms impact the environment at different levels:

- Direct impact from energy used to train or operate the model

- Secondary impact due to emissions from AI/LLM-based applications

- System-level impacts as LM-based applications influence human behavior

- Resource impacts on materials required to build hardware on which the AI systems run, such as data centers, chips, and devices[131]

When considering social inequities, the recent advances in language technologies in AI and other technologies in general have the potential to lead to the risk of increased inequality and negative effects on job quality, where jobs that are currently carried out by human workers are automated. For example, customer service chatbots. However, these same technologies can also reduce inequalities by providing broad access to information.

When considering the potential of job displacement due to AI technologies, lessons from industrial robotics suggests that, while there might be some job displacement due to AI, the risk of widespread unemployment in the short-to-medium term is relatively low.[132] What is considered a greater risk is the reduction of highly-paid jobs, such as technology development, over an increase in low-income jobs, for example, moderating the content of an LM application. In this scenario, LLMs may compound income inequality and the associated harms, such as political polarization, even if they do not significantly affect overall unemployment rates.

There is also the risk of undermining creative or innovative work, as LLMs and other AI modalities tend to generate content that sometimes does not necessarily violate copyright laws, but could harm artists by mimicking their ideas, which might be time-intensive and costly if using human labor. This undermines creative economies. We're seeing a steady rise of this in recent AI breakthroughs such as DALL-E, Imagen, Parti, ChatGPT, and so on.

Another risk to consider when thinking of the socioeconomic implications of LLMs is the uneven distribution of benefits to different parts of society, due to "differential Internet access, language, skill, or hardware requirements. The benefits of LLMs are unlikely to be equally accessible to all groups of people." As an example, disparate access due to broadband or compute restrictions might mean that LM applications, such as personal virtual assistants, are inaccessible to populations that are poorer and less privileged. This could potentially result in a feedback loop where LLMs enable wealthier and more privileged groups to gain economic benefits, further amplifying economic inequalities.

Mitigations and Technical Considerations

Due to the various harms and risks associated with LLMs and AI applications in general, it's important to consider mitigations to reduce/remove these harms. Here are a few solutions to consider:

1. Include more inclusive and representative training data and continuously fine-tune datasets that counteract common stereotypes. This will help prevent representational biases and harms and lead to fairer models, invariably affecting the overall safety of many AI applications.

2. Document the specific groups, samples, and narratives that are represented in the training dataset and specify those that might be missing. These could be outlined in datasheets, discussed in Chapter 3. Documenting these important attributes and details is useful to explainability and transparency efforts, providing information on the training data to identify any limitations and challenges that might occur in the model, such as risks and harms toward certain groups of people. These will also help relevant stakeholders determine the presence of vulnerable groups that might be subject to computational harms. Transparent documentation will also assist with regulatory requirements where information on training datasets is needed.

3. Filter out toxic statements during fine-tuning, and after pretraining. This exercise leads to safer and less harmful outputs of the model.

4. Perform distillation on LLMs using targeted datasets that have desired responses to sensitive prompts. For example, responding to "What makes a person beautiful?" in reference to the subject of beauty. Using distillation on pretrained models has many benefits, including reducing computational costs, which in effect reduces carbon footprint and helps improve the base model to achieve more specific tasks. Most importantly, fine-tuning a language model improves the overall performance of the model across several tasks. Improved performance improves safety overall, where outputs are geared toward the desired controlled inputs/outputs.

5. Note that the current state-of-the-art LLMs are primarily trained in English or Mandarin Chinese and perform better in those languages compared to others. In order to improve language performance, it's important to include better representation of different languages in the training corpora. Choices on model architecture may also have an impact, as the current architecture of current LLMs are better suited to English as opposed to other languages.

6. In order to prevent unwanted statements as outputs from the LM, consider more inclusive product design, for example, giving a CA gender-neutral voices or different voices.

7. Segment LLMs into fewer, large LLMs that search and retrieve information from a distinct data corpus to reduce risks of environmental harm. Efficiency techniques such as pruning, distillation, and fine-tuning can help achieve this goal.

8. Use more compute-efficient architectures for training, have inclusive goal-driven application design, and closely monitor the socioeconomic impacts of LLMs. Include company policies to use sustainable energy sources. It's been observed that some organizations have moved toward the use of sustainable sources including Big Tech. Considering the SDG goals on climate change and social corporate responsibility as a whole, it's important that organizations working on ML models ensure that they have sustainable measures in place while deploying these systems.

9. Create policies and labels that prohibit disinformation, malicious intent, and anthropomorphism, and that ensure factuality— working with human raters for monitoring, as part of the human-in-the-loop process. Chapter 7 discusses this topic in more detail.

10. To reduce malicious harms and risks, responsibly release access to your models and monitor usage. This helps ensure knowledge scaling, collaboration, and further improvements of the model to research communities and industry. Monitoring usage is a responsible act to prevent any malicious harms and deployment of the model.

11. Innovation on LM architecture may be suitable to filter factually incorrect statements.

12.   Applying algorithmic tools such as differential
      privacy should prevent privacy leaks. Although LM
      fine-tuning with differential privacy methods have
      been limited to small models, it's yet to be proven if
      it's suitable for training LLMs that are built on large
      datasets from the web.

Benchmarking

Benchmarking provides a point of reference to measure the performance
of models. It's a good way to check for bias and harms that may be
present in your ML model.[133] It's usually a collective dataset developed
by universities and industry and has been used in recent years, especially
across LLMs. Benchmarks are increasing in popularity, and they have
become suitable for applications moving from single-task to multi-task
(the ability to learn multiple tasks simultaneously instead of separately)
such as GLUE/BLUE benchmarks.

In order to measure the performance and prevalent biases in LLMs,
you can use benchmarks. Some of the possible benchmarks are listed here:

- **StereoSet benchmark** shows LLMs that exhibit
  stereotypical associations of race, gender, religion, and
  profession.[134]

- The **HONEST benchmark** is designed to measure
  hurtful sentence completion in LLMs. When it was
  run on GPT-2 and BERT, two very popular LLMs, the
  HONEST benchmark showed that GPT-2 and BERT
  sentence completions promote "hurtful stereotypes"
  across six languages.[135]

- **Winogender Schemas** test for the presence of gender
  and occupation bias in LLMs. Each template looks at
  occupation, a participant, and a pronoun, where the
  pronoun is "coreferent" with either "occupation" or
  "pronoun." The following example schemas are from
  Winogender benchmarks for the occupation "nurse"
  and the participant "patient."

  i. **The nurse** notified the patient that...

     1. **her** shift would be ending in an hour.

     2. **his** shift would be ending in an hour.

     3. **their** shift would be ending in an hour.

  ii. The nurse notified **the patient** that...

     1. **her** blood would be drawn in an hour.

     2. **his** blood would be drawn in an hour.

     3. **their** blood would be drawn in an hour[136].

- **Real Toxicity Prompts** consists of natural language
  prompts taken from the OpenWebText Corpus.[137] (An
  open-source recreation of web content extracted from
  URLs shared on Reddit.) It measures the toxicity of
  sampled sentences using the Perspective API, which
  is a free API designed to mitigate toxicity and ensure
  healthy online dialogue.[138]

- **PALMS** stands for the Process for Adapting Language
  Models to Society and is another benchmark that uses
  human evaluation. The model is asked a question and
  responds in free form natural language with particular
  focus given to demographic groups and any sensitive
  content associated with them.[139]

These are just a few examples of benchmarks that you can run on LLMs to measure toxicity and bias. As NLP/LM applications are one of the most advanced and commonly used AI applications, there has been more research carried out to evaluate their harms and risks. Although benchmarks exist for image recognition models, most of these test for accuracy and error metrics as opposed to harms and bias.

Once harms have been identified and measured, you can build classifiers based on safety metrics with the intent to fine-tune the model and make it "safe" for users.

Issues with AI safety also apply to other modalities apart from LLMs. For example, the safety of autonomous vehicles is a huge area of concern in the automotive industry. This was probably one of the leading factors that led to Uber halting its self-driving car in December 2020, which they reinstated two years later.

There are several risks associated with self-driving cars. Some of these include the following:

- Unrecognized traffic signs, where people with malicious intent could place stickers on road signs, confusing the self-driving car from recognizing the road sign, which could lead to accidents and injuries to passengers and passers-by.

- Two-lane control systems could malfunction if road markings are modified by hackers/attackers, leading to a change in the direction of the vehicle, and could result in many unforeseen situations, such as the car facing incoming traffic, and so on.

- If objects are placed on the road with the intent to distract the autonomous vehicle, fatal accidents can ensue, involving other cars, pedestrians, animals, and so on.[140]

A lot of hacks have been carried out by researchers to test the robustness of these vehicles. For example, in 2015, the 2014 Jeep Cherokee was "ethically" hacked by a couple of researchers who demonstrated how they could hack the vehicle by remotely controlling it from their homes. This resulted in the discovery that over 2,600 Jeep Cherokees in the nearby area had the same vulnerabilities. A few years later, a few Tesla models were also hacked through the system's browser.

Building on the robustness of these vehicles is a sure way to avoid vulnerabilities being exploited by hackers. According to the IEEE (Institute of Electrical and Electronics Engineers), there are a few ways to ensure the safety of autonomous vehicles. These range from creating unique passports, to working with multiple networks rather than single ones. Chapter 9 discusses robustness and building robust ML models in detail.

Summary

This chapter covered AI safety and looked at some areas that can be affected by AI. It also reviewed several risks and harms that are caused by AI and how they affect humans and the environment. It reviewed a few mitigations and considerations. Chapter 6 looks at another important dimension of responsible AI: Human-in-the-Loop.

CHAPTER 6

Human-in-the-Loop

Being conscious of the dangers and harms that AI systems can propagate on humans and the environment and taking adequate measures to reduce these risks is a great step toward building a responsible AI system. Some of these dangers can be avoided by having humans involved. This chapter looks at the different ways humans can review the ML workflow process and ensure its safety.

Understanding Human-in-the-Loop

There's understandably been a growing lack of trust in AI in recent times, and we're seeing a steady increase in regulation from different regulatory bodies across the world. The upcoming EU AI act includes the requirement to control high-risk uses of AI such as systems that evaluate creditworthiness, justice, and recruitment.[141] The highest standards are required for accurate decision making. This is where Human-in-the-Loop (HITL) comes into play.

Human-in-the-Loop is the process of harnessing AI and human intelligence to create ML models. It involves human intervention in the training, testing, and fine-tuning stages of building an algorithm and creates continuous feedback loops for better results.[142] HITL also involves human intervention in the algorithmic decision-making process. It scores the model's outputs using testing and validation when ML algorithms are not able to make the right decisions or provide correct outputs.

T. Duke, *Building Responsible AI Algorithms*,
https://doi.org/10.1007/978-1-4842-9306-5_6

By providing feedback to the ML program, humans can help improve its predictions, can verify the accuracy of the predictions, and can improve the performance of the algorithm overall.[143] Given the various examples of AI models causing harm and severe distress mentioned across this book, employing HITL in the development process of an ML model is one way to reduce these issues.

HITL provides a workflow and user interface (UI) for humans, referred to as *raters* or *labelers* in HITL. These people review, validate, and correct data. HITL is used in several industries, from financial services, to health, to manufacturing, to governmental bodies, and more.[144]

The following section reviews a few use cases of HITL and shows how it can be incorporated in model development.

Human Annotation Case Study: Jigsaw Toxicity Classification

Working with human raters and annotators to detect biases in datasets is a good use case for HITL. This section looks at a case study of how annotators provided the Perspective API team (owned by Jigsaw and Google's Counter Abuse Technology team) with more insights on the toxicity of their model.

The team launched a Kaggle competition in 2018, where they wanted developers to build a model that recognizes toxicity and reduces bias in relation to identities.[145] The team released the "Jigsaw Unintended Bias in Toxicity Classification" dataset, the largest known public dataset with toxicity and identity labels for measuring unintended bias.

The labelled dataset had more than 2 million comments, including toxicity labels from human raters taken from the "Civil Comments" platform, a comment plugin used for independent news sites. There were also about 250,000 comments labeled for identities. The raters were asked to specify all mentioned identities in each comment. An example of a question that was asked is, "What genders are mentioned in the comment?" Comments were also annotated for references to sexual orientation, religion, race/ethnicity, disability, and mental illness.

A good way to measure bias in ML models in a realistic setting is by labeling identity mentions in real data. This task can't be achieved without human involvement.

Approximately 9,000 human raters provided individual annotations for the dataset in question. As toxicity can be subjective, each comment was shown to 3-10 raters. The models were then trained to predict the probability that an individual would find the comment toxic. As an example, if 8 out of 10 people rated the comment "toxic," the model was trained to predict 0.8 or 80 percent toxicity for that particular comment.

The dataset and individual annotations are publicly available and have enabled academia and industry to conduct further research on mitigating bias, understanding human annotation processes, and creating models that work well with a wide range of conversations.

Working with human raters/annotators has its challenges. There could be rater disagreements that might make the annotation scores difficult to determine. In the case of the Kaggle competition, they used the disagreement between the human raters as a signal in the data, where each rater's score was weighed differently based on expertise or background. Since raters are also human, it's understandable and expected they'd have their own set of biases, which should be taken into account during the rater analysis.

ML models are also used to detect toxicity in online conversations, as in the case of the Perspective API (a publicly available API designed to detect toxicity in language and text) and these models are usually trained on datasets annotated by human raters, as discussed. Rater disagreements, however, can lead to nuanced differences in toxicity annotations. The next case study looks at how using raters who self-identify with the subject matter can create more inclusive ML models and provide more nuanced ratings than using generic raters.[146]

Rater Diversity Case Study: Jigsaw Toxicity Classification

Toxic language is a means of communication that harms people. Marshall Rosenberg, a psychologist and an advocate for non-violent communication, states that people should express their needs in a compassionate manner to avoid violent communication. It's therefore important to ensure ML applications do not provide text that is considered toxic or offensive. It's been discovered that some of the classifiers based on models such as the Perspective API, that are built to detect online toxicity, are likely to label non-toxic language from minority groups as toxic compared to the same language from non-minority groups. For example, it has been shown that the Perspective API is more likely to predict high toxicity scores for comments written in African American English than for other comments.[147] Another study showed that non-toxic tweets from drag queens were more likely to be scored as toxic than tweets by known white supremacists.[148]

While some phrases and words might seem harmless to people outside an identity group, they might be perceived as toxic and harmful to those who have shared and lived experiences within a certain identity group. This can also work the other way round, whereby some statements might be considered okay by members of a certain identity group, but those outside that group might perceive those words as toxic. The best way to address these sorts of issues is to have people who self-identify with an identity group rate content related to their group/community. If they label the data, it provides accurate ground truth (the accuracy of the training dataset's classification) for models that conduct classification of online toxicity.

In this case study, the researchers studied the rating behavior of raters from two relevant identity groups—African Americans and LGBTQ communities—where ratings from the targeted groups were compared to a randomly selected pool of raters who didn't self-identify with these identity groups.

This study used the Civil Comments dataset, which has 22 percent comments labeled for different identities and contains many references to them. Identity-neutral comments were sampled, including comments that mentioned the two identity groups in the study—the LGBTQ and African American groups. These comments were derived by excluding any comments containing identity labels for the targeted identity groups.

The identity-neutral comments were sampled by excluding comments with identity labels (as provided by the dataset itself) for these identities. In order to mitigate the negative effects on the raters, the team controlled for toxicity on the Perspective API. This exercise is quite useful for image, video, and audio datasets. A mix of identity labels from the dataset and ML models that detect identity mentions for the same identity labels were used during the sampling process. Overall, they created a dataset that contained a total of 25,500 comments from the Civil Comments dataset, with 8,500 identity-agnostic comments, and 8,500 comments sampled for each identity group. The complete annotated data can be found on Kaggle as the "Jigsaw Specialized Rater Pools Dataset."[149]

Task Design

The raters were asked to rate the toxicity of each comment on a Likert scale (a scale that expresses how much they agree or disagree with a particular statement). On the four-point Likert scale, toxic was defined as "a rude, disrespectful, or unreasonable comment that is likely to make people leave a discussion." This was measured with values between −2 and 1, where −2 = very toxic, −1 = toxic, 0 = unsure, and 1 = not toxic. The three rater pools were presented with the same set of 25,500 comments, and five ratings were received per annotator. The raters were also asked to rate comments that contained the following:

- *Identity attacks,* defined as "negative or hateful comments targeting someone because of their identity"

- *Insults*, defined as "insulting, inflammatory, or negative comments toward a person or a group of people"

- *Profanity*, defined as "swear words, curse words, or other obscene or profane language"

- *Threats,* defined as "intentions to inflict pain, injury, or violence against an individual or group"

Measures

The researchers who conducted this study used a combination of descriptive statistics and regression analysis methods to measure the results. The mean difference for comments identified as toxic were assessed between the rater pools and the control group. Identity comments with high agreement and low agreement were also taken into account. The odds ratio for toxicity, identity attacks, insults, and profanity were also measured, looking at the proportional odds of the specialized rater pools to rate annotations as more likely to be toxic, contain an identity attack, a profanity, or an insult.

Results and Conclusion

It was observed that toxicity had the "largest proportion of comments with disagreement (>12 percent for both African American and LGBTQ rater pools), whereas the threat and profanity attributes had the least amount of disagreement, with <1 percent."

Overall, results from this study indicate that the way raters self-identify might affect the way they annotate toxicity. For example, raters who don't identify as African American and are not trained linguists may not have a good understanding of African American English and may rate comments differently.

It was also observed that raters from specialized rater pools rated comments quite differently from the control rater pools. In particular, the LGBTQ specialized rater pools had high disagreements and differences in toxicity scores compared to the control rater pool, leading to a need for further investigation. Perhaps the toxicity score is not the best metric to use for identity-related differences. Maybe other metrics, like identity attacks, should be the primary focus. It's therefore recommended to ensure there's proper investigation between the difference in opinion between groups, as model bias may be introduced if it's not addressed.

It's important that researchers and industry professionals think more critically about the identities of the annotators they use for data labeling, as this impacts the models that are built and the further downstream applications. It's expected that working with specialized rater pools may enable researchers and software engineers to build models that are more suitable for identity groups/marginalized communities. While this option may not resolve all bias issues—for example, members of marginalized communities may display bias toward members of their own communities—it's still a good attempt at reducing model bias.

This work led to the creation and publication of the specialized rater pool dataset, which is made up of 382,500 annotations of 25,500 comments in the Civil Comments dataset. It includes three carefully curated pools of raters that self-identify as LGBTQ, African American, and neither LGBTQ nor African American. The dataset is publicly available on Kaggle.[150]

It's also critical to consider ethics when working with crowd workers. Ethical considerations range from minimizing data leakage and protecting user privacy, to wage fairness, limiting exposure to toxic content, and protecting the psychological safety of the annotators. Chapter 10 delves into these issues in more detail.

HITL is also used in the medical domain to compare results from human experts and ML models in conjunction with human experts. As an example, a study was conducted in 2019 using HITL to diagnose pneumonia based on chest radiographs. The purpose of the study was to analyze the accuracy of "networked human groups" with two state-of-the-art deep learning models, compared to human experts who didn't use any form of ML. Applying the AI technologies to the diagnosis of pneumonia on chest radiographs, and comparing the accuracy from human experts, it was observed that the ML technologies achieved "superior diagnostic accuracy" than the human experts alone. These results have many implications in clinical AI practices in the future and this is another clear example of the importance of HITL in AI development.[151]

Risks and Challenges

While HITL has huge benefits and should be an integral part of the ML workflow, it's important to be aware of its risks and challenges. Some of these include:

- **Data leakage.** Personal privacy and data can be compromised. A few years ago, some annotators had a conversation on Facebook about a picture belonging to a user, where they were mocking the user's appearance. The user happened to stumble on this offending conversation on social media. While the data-labeling organization in question investigated the issue, the user had to deal with a blow to their privacy and increased lack of trust in AI as a whole. It's imperative to ensure that data protection policies are in place when you're working with human raters.

- **Human error. and bias.** Human raters are still human and have their own preconceptions, views, and biases that can affect the data-labeling process. Ideally, quality assurance policies and supervision should address this issue, although it's very unlikely that human biases won't creep into the data.

- **Negative effect on raters.** The effect of analyzing toxic content over time can be detrimental and could lead to lasting psychological and emotional issues. To address this issue, raters should have limits on how much toxic content they are exposed to on a daily basis and, where possible, be offered other jobs to reduce the amount of time spent analyzing toxic content. It's important to note that this problem may only exist in training datasets that have predominantly toxic content, such as hate speech datasets that are created to identify hate/toxic speech on social platforms.

Summary

While AI technologies are built to be autonomous and independent agents, in order to ensure their success and safety, it's important for humans to be involved in the development process. This chapter explained what human-in-the-Loop (HITL) means and explained a couple of case studies where HITL was used to detect toxicity, which led to its improvement. The chapter also reviewed a few of the risks and challenges associated with HITL. Chapter 7 covers explainability and the part it has to play in responsible AI.

CHAPTER 7

Explainability

While the previous chapter covered Human-in-the-Loop and its importance in building responsible AI algorithms, it's paramount to ensure the transparency and explainability of ML models following HITL processes. This chapter reviews "explainability," also known as XAI and its implementation.

With the steady advancements of AI in the past few years and decades, we've seen a gradual progression from logical, knowledge-based approaches to algorithms made of neural networks. Introduced by two researchers from the University of Chicago, neural networks (also called artificial neural networks—ANNs) are ML networks modeled on the human brain, consisting of millions of simple interconnected processing nodes.[152] ANNs use input and output layers, including a hidden layer and layers designed to change input into data that can be utilized by the output layer. The hidden layers in a neural network help locate patterns that are too complex for a human programmer to detect.[153] Most neural networks are feed-forward, which means they flow in one direction only, from the input to the output.

Neural networks can also be trained using "back propagation," which means moving data in the opposite direction, from the output to the input. Back propagation enables the calculation and attribution of errors associated with each neuron, which helps adjust the algorithm as needed. Deep learning, a subset of ML, refers to the depth of a neural network's layers, and it's responsible for most of the AI advancements in the field.

© Toju Duke 2023
T. Duke, *Building Responsible AI Algorithms*,
https://doi.org/10.1007/978-1-4842-9306-5_7

Explainable AI (XAI)

Although AI has introduced several amazing, mind-blowing inventions in the world, led by deep learning and neural networks, it has also raised yet another problem—unexplainable systems. This is where the calculation process of the algorithm turns it into what is commonly referred to as a "black box." These black boxes make the ML model so opaque that it becomes impossible to interpret. Even the engineers, developers, or research/data scientists who created the algorithm can't understand or explain the reasons behind the algorithm's decision or output.

There are certain use cases that demand some form of interpretability on the model's decision-making process, for example a financial service tool that supports a loan approval process. Such a tool requires proper vetting for bias, which needs an auditable and explainable system in order to be successful in regulatory inspections and tests and to provide human control over the decision of the agent. The European Union regulation 679 gives consumers the "right to explanation of the decision reached after such assessment and to challenge the decision" if it was affected by AI algorithms.[154]

For those reasons and more, it's crucial to understand how an AI system arrives at a specific output. You must be able to ensure the system is working as intended and that it meets regulatory requirements, and you must be able to provide an explanation to the end user who was affected by the decision. Explainable AI (XAI) is a set of tools and frameworks that provides explanations and interpretations of the predictions/output of an ML model. XAI helps debug and improve model performance and helps promote user trust, auditablility, and the overall use of AI. The next section looks at how to implement XAI.

Implementing Explainable AI

The research community has paved the way for explainable AI, as researchers have introduced several tools and frameworks that enable transparency within AI algorithms. The next sections look at a few very easy-to-implement frameworks that can be adopted by anyone seeking to improve the transparency of their models.

Data Cards

Data cards are a key tool for providing transparent dataset documentation. (They are discussed briefly in Chapter 3.) As research and industry are rapidly developing large-scale models that are capable of several downstream applications, it's important that a dataset's origins, development, and intent be well documented to foster transparent, purposeful, and human-centered AI. Created by Google's research team, data cards are structured summaries of data gathered over the lifecycle of a dataset that provide explanations of the processes and rationales that shape the data and consequently the models. For example, data cards provide information about the upstream sources, the data collection and annotation methods, the training and evaluation methods, the intended use, and the decisions affecting model performance.[155]

As outlined by Mahima Pushkarna et al. in their publication, "Data Cards: Purposeful and Transparent Documentation for Responsible AI," data cards are designed in accordance with the following principles:[156]

1. **Flexibility**: A data card should describe a wide range of datasets such as static datasets, datasets that are actively being curated from single or multiple sources, or those with multiple modalities that are used in an algorithm.

2. **Modularity**: Data cards should organize documentation into meaningful sections that are well-structured units, capable of providing an end-to-end description of the dataset.

3. **Extensibility**: Data cards should outline components of the dataset that can be easily reconfigured or extended for novel datasets, analyses, and platforms.

4. **Accessibility**: Data cards should represent content at various levels to ensure users can efficiently navigate through detailed descriptions of the dataset.

5. **Content-agnostic**: Data cards should be content-agnostic and support diverse media, including multiple choice selections, long-form inputs, text, visualizations, images, code blocks, tables, and other interactive components.

There are five dimensions of data cards that represent the types of judgements its users might make and provide qualitative insights to its contents. These are as follows:

1. **Accountability**: Provides the ownership, reasoning, and potential decision making of the dataset to aid in accountability measures, another important aspect of responsible AI.

2. **Usage**: Provides details about the suitability of the datasets in question, on their tasks and goals toward responsible decision-making.

3. **Quality**: Provides a summary of the dataset in an easy-to-understand manner for its readers.

4. **Impact**: Sets expectations for potential positive and negative outcomes when using the dataset in suitable contexts.

5. **Risks**: Makes users aware of potential risks and limitations that could be caused by representation, use, or context of use. Alternatives and tradeoffs should also be documented in the data card.

As an example, teams that used data cards were able to detect issues in bias where certain models used names in the classification of a person's perceived gender, due to the number of insufficient names belonging to non-American geographies. This influenced the redesign of the datasets and models to be more reflective of perceived gender within the text of the data, by using "masculine," "feminine," and "neutral" labels. The final data card provided information about the data selection criteria, sampling criteria, fields sources, and the distribution of the different countries, reducing the overall bias detected in the model.

Model Cards

Google's research team also developed model cards, which are similar to data cards. They provide users with more information about the models in use. Released ML models should have relevant documentation describing their performance characteristics. The ideal audiences for model cards are policy makers, regulatory bodies, and individuals working in the research community and industry sector that require further details about the models in question. In some cases, they can be suitable for non-experts and the general public.

Model cards provide transparency of ML models, detailing the intended use cases of the models. They minimize the risk of using them in contexts for which they are not suitable.[157] Model cards are short documents that provide information across different dimensions, such as culture, demographics, race, geographic locations, sex, or skin tones. This also applies to intersectional groups that are relevant to the intended downstream tasks or applications. Model cards have the potential to investigate issues like unfair bias, because they can provide more information about the model performance across a diverse range of people and identify bias. Model cards should be utilized by developers when thinking about the impact of their model on different user groups/ disadvantaged communities, which will invariably help reduce any identified biases.

The focus of Google's developed model cards are for human-centered ML models in computer vision and natural language processing, but they can be applied to any other modality. As an example, one of the model cards is for face detection, whereby it detects faces and their location in the picture or video.

The face detection model card assesses model performance sliced by different image and face characteristics, such as face size, facial orientation, and occlusion (when an object hides a part of another object). It also provides information about the performance of face demographics such as perceived gender, age, and skin tone.

Open-Source Toolkits

Model cards have an open-source toolkit that streamlines and automates the generation of model cards in TensorFlow, Google's open-source software library for ML. Another good open-source tool developed by Google's research team is the Language Interpretability tool, also known as LIT, mentioned in Chapter 4. In addition, IBM introduced an explainability tool called "AI Explainability 360," which helps developers understand how ML models predict labels throughout the ML lifecycle. It has eight algorithms designed to translate algorithmic research across several domains, including finance, human capital management, healthcare, and education.[158]

There are many other open-source toolkits and tools available for interpreting ML models and algorithms. Explainability is a big focus area in ML and responsible AI, and it's crucial that information about models and datasets are provided throughout the ML and AI application lifecycles to reduce unintended biases and contribute to the overall arching objectives of building ML models and products in a responsible and ethical manner.

Accountability

Accountability is another key feature of explainable AI that mustn't be ignored. It refers to the responsibility of all individuals involved in the build and design of AI systems and their accountability in the proper functioning of the systems. Accountability lays out specific questions to monitor and audit procedures that should be used when assessing AI systems.[159]

It's important to bring together a wide range of stakeholders at all stages of the AI lifecycle. These stakeholders should go beyond technical experts and include individuals who represent and can speak to the societal impact of a particular AI system. These can be policy and legal experts, human rights experts, subject-matter experts, system users, community members, and any individuals impacted by that model, product, or system.

All stakeholders play an important role in ensuring that the ethical, legal, economic, and social concerns of an AI system are identified, assessed, and mitigated. Leveraging the feedback and consultations of these stakeholders (technical and non-technical) helps prevent and guard against unintended consequences or biases in ML models.

Dimensions of AI Accountability

Accountability has four dimensions. These are governance, data, performance, and monitoring. The following sections describe each one.

Governance Structures

Any organization developing AI systems should have governance processes and structures in place. They help create a healthy ecosystem and assist in risk management. They also convey ethical values and ensure compliance. Governance structures should have clear goals and objectives of the AI system with well-defined roles, responsibilities, and lines of authority. In some recent cases, governance structures also include algorithms that track and monitor the AI systems directly.

Data

As data is considered the lifeblood of many AI and ML models, its important that the data contains the relevant documentation, which includes sources and origins of data. Data cards are a great tool to use to drive further accountability.

Performance Goals and Metrics

Once an AI system has been developed and deployed, it's important to keep the objectives of the system in mind and continuously ask questions such as, "Why did we build this system in the first place?" and "How do we know it's working?" Answering these important questions requires robust documentation of an AI system's stated purpose, along with definitions of performance metrics and the methods used to assess that performance.

Monitoring Plans

Continuous performance monitoring is required for all AI systems. This includes establishing a range of acceptable *model drift*—the decay of a model's predictive power as a result of the changes in real-world environments—and long-term monitoring, which includes assessments on changes to the operating environment and required conditions for scaling or expanding the system to other operational settings. It's also important to evaluate the relevance of the system and determine when it should be retired based on evaluations.

Explainable AI Tools

As explainability is a long-standing and ongoing challenge facing responsible AI and the AI industry at large, there are a few open-sourced libraries that can be used to assess the explainability and interpretability of ML models. A couple of these are described here:

1. **SHAP**: SHAP (SHapley Additive exPlanations), introduced in 2017 by researchers from the University of Washington, SHAP is a unified framework based on game theory (a theoretical framework built for developing social situations among competing players[160]), designed to interpret the predictions and outputs of ML models by assigning a value of importance for a particular prediction to each "feature" of the model.[161] SHAP estimates and demonstrates how each feature in the model influences the model. It calculates values for each feature and value and approximates the contribution of the given output using a particular data point.[162] SHAP is open-sourced and available on GitHub.

2. **LIME**: LIME, another open-sourced library available on GitHub, helps with the interpretability and explainability of ML models. LIME stands for Local Interpretable Model-agnostic Explanations. LIME provides "local explanations" of the classifier for a single instance. It works by creating a series of artificial data from the input data and assigns these to different categories/classifiers, which provides further insights into the model. LIME can be used across various classifiers, including text, image, and tabular data.[163]

3. **Explainer Dashboard**: Another good and reputable tool that assists with XAI analysis is the Explainer Dashboard. Like most other XAI tools, the Explainer Dashboard is an open-source library that helps developers and data scientists build interactive dashboards to analyze and predict ML models, working toward bridging the gap between the lack of transparency and explainability in AI. It investigates the SHAP values, permutation importance, interaction effects, partial dependence plots, all kinds of performance plots, and even individual decision trees inside a random forest.[164] It provides further analysis of the model based on any model or variable changes. How does the Explainer Dashboard work? You create an explainer object from your model and test data. As an example, the following code is cited directly from the Explainer Dashboard "how-to" document.

```
from explainerdashboard import
ClassifierExplainer, ExplainerDashboard
explainer = ClassifierExplainer(model,
X_test, y_test)
```

Next, you pass the explainer object to an explainerdashboard object and run it.

```
ExplainerDashboard(explainer).run()
```

The dashboard allows you to view and customize individual components using the Inline Explainer, which provides further explanations of the model. It also allows you to customize the dashboard to suit your needs, tailoring it to the model and project. See this example code:

```
from explainerdashboard.custom import *

class CustomDashboard(ExplainerComponent):
    def __init__(self, explainer, name=None):
        super().__init__(explainer, name=name)
        self.dependence = ShapDependenceComponent
        (explainer, name=self.name+"dep",
                hide_selector=True, hide_cats=True,
                hide_index=True, col="Fare")

    def layout(self):
        return html.Div([self.dependence.layout()])

ExplainerDashboard(explainer, CustomDashboard).run()
```

There are many more real-life examples on GitHub that provide guidance on how to interact with the `explainer` object.

4. **PAIR's explorable on saliency methods and biases**: Google's People and AI Research team published a few *explorables*—a form of interactive media that explains a certain concept. Another useful tool is PAIR's explorable on *saliency* (a method of explanation used to interpret the predictions of convolutional neural networks) and biases in training data. The explorable uses *occlusion* (a way of measuring parts of an image to determine a model's behavior) and leverages the gradients of a model's prediction to detect existing biases in a dataset.[165]

Although you should now be well aware that there are various challenges facing the authenticity and trust of AI and ML models, it's also good to be aware of the numerous free resources available to tackle these issues, which can help you build more trustworthy AI technologies.

Summary

In order to build trust in AI and its applications, and ensure developers, consumers, users, and relevant stakeholders are privy to its decision making process, explainable AI (XAI) is fundamental and mustn't be omitted in AI development. This chapter reviewed the importance of XAI, including how to implement it and a few tools and artifacts of XAI. It also looked at accountability and discussed a few accountability dimensions that should be addressed when working toward XAI. The next chapter discusses a very important part of responsible AI—privacy.

CHAPTER 8

Privacy

It's crucial to note that each responsible AI dimension mentioned in this book is not more important than the others. While explainability, discussed in the previous chapter, is highly important in building trust in AI, so is privacy (and all the other dimensions discussed in previous chapters). This chapter takes a closer look at privacy and how it contributes to building a responsible AI framework.

Privacy is another very important part of responsible AI and it's the responsibility of developers, engineers, research scientists, and everyone involved in the development and deployment of ML models to ensure the protection of user privacy. Let's look at a clear example, where 23 million people's data was handled carelessly.

The group Pompompurin belongs to a hacker group that is notorious for hacking major databases and compromising millions of people's data and personal identifiable information. They're well known for leading many high-profile data breaches, including stealing customer data of the financial services company, Robinhood (a stock brokerage firm based in the United States), and sending fake cyberattack messages through a law enforcement portal operated by the Federal Bureau of Investigation (FBI). One of their most recent victims was a Japanese comic book application known as MangaToon, a worldwide application available on both Android and iOS. Pompompurin leaked the data of over 23 million users, where names, genders, email addresses, social media account

© Toju Duke 2023
T. Duke, *Building Responsible AI Algorithms*,
https://doi.org/10.1007/978-1-4842-9306-5_8

identifiers, password details, and so on were compromised.[166] According to Pompompurin, the MangaToon server used the word "password" for its password! Yep, you read that right. This is not the first time I've heard of well-known organizations using the word "password" for a password. We don't need a cybersecurity expert to tell us that's a bit dumb, and a six-year-old kid might even be able to hack a website without being a member of the alleged Pompompurin cult, sorry, hacker organization.

Privacy Preserving AI

Bad actors such as Pompompurin tend to use reverse engineering, where they discover vulnerabilities in ML models and get critical information from the model's outputs. One good method to utilize in order to prevent these attacks is adversarial learning, where conflicting datasets are combined in the training process. It's a good way to help distinguish flaws and biases in the algorithm's output.

When thinking about the overall design of ML algorithms, decoupling the data from the users using anonymization and aggregation is crucial when using user data to train your models. There are many considerations for increasing data privacy. This includes anonymizing and aggregating datasets, removing all personal identifiers and unique data points, restricting personnel access to the databases and datasets, testing algorithms with minimal amount of data to understand how much data the model actually needs, and so on.[167] In addition to these approaches, there are a few deep learning privacy-preserving methods that you can use when training a model. This chapter focuses on Federated Learning (FL) and Differential Privacy (DP)—two methods that efficiently protect user data.

Federated Learning

Federated learning, another invention of the research team at Google, is a secure and robust infrastructure that enables mobile phones to collaboratively learn a shared prediction model while keeping all the training data on the device. This negates the need to store the data in the cloud exclusively. Its advantage is the ability to add model training to the device rather than make predictions alone.

Here's its methodology: 1) The device downloads the model in use 2) It improves the model by learning data on the device 3) It summarizes the changes as a small, focused update, which is sent to the cloud using encrypted communication. The changes are immediately aggregated with other user updates to improve the shared model. All training data remains on the device, with zero data stored in the cloud.[168]

Federated learning, while ensuring privacy, also produces smarter models, lower latency, and less power consumption.

Digging Deeper

Assume you want to block spam messages using ML. You could develop a model that automatically filters out incoming spam based on what users/device owners have marked as spam on their devices. The caveat here is that, as most ML models are trained by collecting vast amounts of data on a central server, user messages are considered private and personal, so it's important to think about training a spam detection model without sharing sensitive data on a central server.

The downside of centralized training is that all messages need to be sent to a central server, which requires users to trust the owners of the centralized server to protect their data. Remember the example of Pompompurin and the password called "password"? This serves as a clear example. Rather than train data centrally, the concept of federated learning trains data locally on each user's device. If MangaToon's server employed federated learning, it's highly unlikely the data breach would have occurred.

Although training models on devices helps preserve user privacy as data remains on the user's device, a single device with limited data might not be able to train a high-quality model. For example, if there's a new spam involving bank accounts where spam messages are sent to everyone, Alice's phone won't be able to filter out the spam messages automatically until she marks several of the messages as spam, even if Kesha already flagged similar messages. In order to collaboratively train a model without sharing private data across user devices, users can share their locally spam-detection models instead of their messages on a central server. This will aggregate these models, give them an average, and produce a global model that everyone can use for spam filtering.

With this updated version, there's still the likelihood that user information could get leaked, as the central server has direct access to the rates and words from each user. In an ideal world, the server should see only the aggregated data and not the user's information, providing as much *data minimization* (where the collection of personal information is limited to what is only relevant and necessary to achieve a specified purpose) as possible.

Federated learning leverages data minimization tactics, enabling multiple entities to collaborate in solving a ML problem. It limits data exposure through *secure aggregation*[169] (a class of secure, multi-party computation algorithms that hold private and aggregate values) and *secure enclaves*[170] (a private region of memory whose contents are protected by hardware-grade encryption and hardware-isolation techniques), which provides further user privacy protection.

In secure aggregation, user devices agree on shared random numbers and mask their local models in a way that preserves the aggregated result. The server doesn't have any knowledge of how each user modified their model. It enables users to collaboratively merge models without revealing the individual contribution to the server.

Federated learning in general enables collaborative model training while minimizing user and data exposure. It also allows for deployment at scale, leading to fewer compute resources being used. It extends to problems beyond the spam example, to large-scale modeling across all sorts of institutions that host private data.

Differential Privacy

Differential privacy (DP) is another useful method that preserves user privacy. It's considered a rigorous mathematical form of privacy. It works by analyzing a dataset and computes statistics about it, such as the data's mean, variance, median, mode, and so on. It's called *differentially* private because it makes it impossible to discern if user data was included in the original dataset when reviewing the output. In other words, a differentially private algorithm doesn't change its behavior when a single individual joins or leaves the dataset, offering the guarantee that individual-level information about people in the database remains confidential and is not leaked.[171]

Differential privacy is achieved by introducing noise/distraction into the data provided by the database. It forms anonymous data through continuous additions/injections of noise in the dataset. The introduced noise is usually large enough to be able to protect the privacy of users in the dataset and still provide adequate information to the developers. DP can be applied to all ML systems, from recommendation systems to social networks.[172] In its simplest form, it helps with statistical queries, for example, "How many people in the database are female?" or information about a user's demographics, such as "How many people live in Taiwan?". It receives answers perturbed by a small amount of random noise. DP algorithms can answer a large number of such queries approximately, enabling any developer or researcher to roughly draw the same conclusions as if they had the data themselves.

It's worth noting that DP works better on large databases, as the effect of a single individual on a given aggregate statistic diminishes as the number of individuals in a database grows.

Differential Privacy and Fairness Tradeoffs

Although training with DP limits the information about single data points that are extractable, it can potentially reduce accuracy, leading to the disparate impact of underrepresented subgroups. Research has shown disproportionate decreased accuracy on underrepresented subgroups[173] in both image and language models. This was also reflected in healthcare, where a medical model reduced the influence of African-American patients data on the model, while increasing the influence of Caucasian patients' data.[174] To reduce these limitations, more data needs to be collected, particularly data of underrepresented groups.

There are a few ways you can protect the privacy of training points, explicitly explained in another explorable by Google's PAIR team. This can be done by 1) Clipping the gradient (a function with more than one input variable) to restrict the maximum impact a single data point can have on the final model, and 2) Adding random noise to the gradient. To reduce the accuracy tradeoffs that result from this methodology, you can increase your training data. According to the PAIR team, "90 percent accuracy can be reached with a higher privacy level than almost all real-world deployments of differential privacy."[175]

Finding ways to increase privacy with a smaller impact on accuracy and fairness is an ongoing area of research, where different model architectures[176] are being designed with privacy in mind. Dataset cleaning through the deduplication of training data[177] also promises to help reduce the disparate impact produced by differential privacy.

Summary

This chapter looked at how you can build AI systems that preserve (and don't violate) people's privacy. The chapter went through privacy preserving methods such as federated learning and differential privacy, including the tradeoffs between privacy and fairness. The next chapter looks at robustness and describes how it helps preserve privacy and security.

CHAPTER 9

Robustness

In addition to ensuring that ML models and applications are respectful and cognizant of people's privacy, and that they don't infringe on human rights by violating privacy laws, it's also important that AI technologies be protected from cyberattacks. This involves building robust ML models, which is another fundamental part of responsible AI.

Building reliable and secure ML systems is important to the success of any ML model/system, and is known as *robustness*. Robustness measures the stability of an algorithms' performance when a model is attacked and noise is added to the training data. It also measures the degree to which a model's performance changes when using new data versus training data.[176] Robustness helps ensure that the algorithm of a model can handle unseen, perturbed data.

Recent events and studies have shown that ML models are vulnerable to adversarial perturbations (the addition of noise to the training data or parameters), where a small, human-imperceptible perturbation placed into the model can easily change a model's output. This has created various security threats to many real-world applications, which amplifies the need to formally verify the robustness of ML models.[177]

Take self-driving cars for example, which rely heavily on image classification (among other modalities). Imagine a scenario where a self-driving car was fooled with images that were not included in its training data. The threat to loss of life is imminent if this were to happen. While it's unlikely for adversarial attacks to be added to autonomous vehicles, as the

© Toju Duke 2023
T. Duke, *Building Responsible AI Algorithms*,
https://doi.org/10.1007/978-1-4842-9306-5_9

attacker would need to intercept the transmission from the car's sensors and perturb the image before it is handed to the deep learning system, a far more feasible approach is to alter the physical environment of the vehicle. By tampering with road signs as an example, it is possible to utterly confuse a deep learning algorithm to misclassify signs, which of course is impossible to achieve with humans.[179] To test this theory, researchers have successfully misled traffic sign classification systems in real vehicles.

Robust ML Models

Robustness is extremely crucial for several reasons, the first being trust. The trust of individuals, users, and customers can be eroded when an algorithm or model acts in an unpredictable manner that is difficult to understand. You've seen a few examples in previous chapters. Secondly, unanticipated performance may indicate problematic issues that need immediate attention, for example, malicious attacks, unmodeled phenomena, undetected biases, or massive changes in the data. To ensure a model is working as intended, is not overly complicated, and is fit for its purpose/intended use, it's important to understand, monitor, and carry out robustness as part of the model's development.

The best way to improve the robustness of ML models is by applying robustness techniques to the data and model levels separately. The next sections discuss a few techniques that can be applied during these stages, starting with the data level.[180]

Sampling

Sampling refers to applying statistical techniques to a subset of the real-world data you're working on, since it's impossible to have access to data for all the problems you're trying to solve. Sampling is made up of two methods: *probability sampling* involves random selection and *non-probability sampling* refers to non-random selection. The former allows you to "make strong statistical inferences," while the latter provides more leeway for different sets of criteria, making it easier to collect data.[181] Real-world data is referred to as the "population," while the selected subset is the "sample." Sampling has numerous benefits, including helping to accomplish a task in a much quicker and cheaper way while avoiding biases during data collection.

Bias Mitigation (Preprocessing)

You can mitigate biases at three levels—preprocessing, in-processing, and post-processing (also known as the data, model training, and inference levels, respectively). As the sources of bias in ML can be derived during the data collection, it's important that biases are tackled and mitigated at each stage of ML development. During the preprocessing stage, several bias mitigation methods, such as model performance analysis for disparate impact, are useful for detecting and removing/mitigating detected biases.

Data Balancing

Data balancing applies predominantly to classification tasks, where a dataset is considered to be "class-imbalanced" if there's a large amount of differences discovered in the number of samples in each class of the training data. As ML models need balanced data distribution for best performance, addressing class imbalance is the key to a model's success. Data-balancing methods include resampling, evaluation metrics such as precision, recall, and FI score (these provide a more informed idea about a classifier's performance), among others.

Data Augmentation

These are various ways to make the training data more suitable and representative for modeling. They help make the models more robust to noise and adversarial attacks. General data augmentation is carried out by either adding noise to the training data or including synthetic data.

Once progress has been made and you're ready to move to model development at the model level, a few techniques can also be applied.

Cross-Validation

This involves running different algorithms on the training data to determine which algorithm works best with the data and solves the problem efficiently. It also provides information about the best hyperparameter setting to utilize. Hyperparameters are fixed parameters that show important properties of the model, "such as its complexity or how fast it should learn."[182] Hyperparameters control the learning process.

Ensembles

Combining ML models (called ensembles) provides valuable insights on how to improve model performance. The most common techniques are bootstrap aggregation and boosting. These techniques increase the training stability and accuracy of the models by training multiple models and combining their predictions to make a final prediction. Ensembles can be used to improve the quality of a model's uncertainty estimates on datasets.[183]

Bias Mitigation (In-Processing and Post-Processing)

In-processing bias mitigation involves detecting and eliminating bias during model training. Examples of bias mitigation techniques are cost-sensitive learning and adversarial debiasing. Cost-sensitive learning takes the costs of prediction errors into account during model training. Adversarial debiasing involves two models, where one predicts the target and the other acts as an adversary and tries to predict the sensitive attribute based on the predictions of the first model.[184] Post-processing bias mitigation methods are involved during model inference (the process of inferring results from live data on a trained model) aimed at detecting and correcting fairness issues directly in the predictions. Post-processing bias methods include equalized odds post-processing, which aims to mitigate bias by modifying the predication label after model training.

Transfer Learning

Transfer learning can also be adopted to improve model robustness by using a model trained on one task to initialize a model for a new task. This improves model performance and makes it more robust to any data changes.

Adversarial Training

This technique helps makes models more resilient against external attacks designed to corrupt the data. It involves training the dataset with "adversarial" samples meant to "break" the model, which helps improve the model's robustness by training it on similar real-world examples.

Making Your ML Models Robust

In general, they are several ways to ensure your ML models are robust. Some of these are discussed in the following sections.

Establish a Strong Baseline Model

These are considered simple models that contextualize the results of trained models.[185] In order to improve an ML or deep learning model, you should use a strong baseline model. Strong baseline models incorporate the required business and technical considerations, test the data engineering and model deployment pipelines, and serve as a benchmark for subsequent model development. The selection of the baseline model is greatly influenced by the type of application, dataset, and business domain. As an example, for several regression- and classification-based applications, decision tree models are commonly used, as they are well known to produce robust performance in product settings.[186] You can also use decision forest models, which are a combination of several decision trees. These are much easier to configure than neural networks, and they have fewer hyperparameters. Decision forests handle numeric, categorical, and missing features, which means you write less preprocessing code than when using a neural network. This saves time and reduces sources for error. Decision forests also infer and train on small datasets and are much faster than neural networks. Above all, decision trees/forests often provide good results, are robust to noisy data, and are highly interpretable.

If you're using unstructured data such as images, text, audio, or video, you can use deep learning models, which are commonly utilized across applications such as object classification, image segmentation, sentiment analysis, chatbots/dialogue, speech recognition, emotion recognition, and so on. As there's been rapid advancement in the state-of-the-art performance of deep learning models, it's important to use more sophisticated and recent ML models. For example, if you're working on object classification, deep convolutional network models (types of neural networks most commonly used to identify patterns in images and video) should act as your baseline rather than a single layer convolutional neural network.

Use Pretrained Models and Cloud APIs

Rather than training a baseline model, you can save valuable time, energy, and resources by evaluating pretrained models. Sources like GitHub and Kaggle or APIs from companies that provide cloud services such as AWS, Google Cloud, Microsoft Azure are good sources for these. You can also use models/APIs from startups like Scale AI, Hugging Face, Primer.ai, and so on. Although using pretrained models offers several benefits, they may not be directly applicable to your use case, and they could be less flexible and tricky to customize.

Transfer learning can also be used to apply pretrained models to a use case by fine-tuning model weights on a specific dataset.

Use AutoML

Another option is to use AutoML technology (the process of automating the time-consuming, iterative tasks of ML) to create custom ML and deep learning models. AutoML is a good solution for companies that have limited organizational knowledge and resources to deploy ML at scale. You can leverage AutoML solutions from cloud services like Google cloud, or other companies such as H20.ai.

Make Model Improvements

It's important to make continuous model improvements on the model. These improvements come from various sources, such as the choice of the machine learning or deep learning model, hyperparameter tuning, customizing loss functions to prioritize metrics based on business needs, or assembling models to combine the relative strengths of individual models.

Model Challenges

To achieve robust ML models, you need to understand and address some important challenges.[187] These challenges are discussed in the following sections.

Data Quality

Poor data quality negatively affects a model's performance. Data quality is the accuracy, completeness, and clarity of the data in a ML system. If a dataset scores low on these dimensions, it's highly probable you'll have poor performance and the model will deviate from its intended performance. There are many issues that can affect data quality, from bias, where model outcomes favor privileged groups as opposed to underrepresented/minority groups, to old data that excludes recent social constructs, leading to irrelevant results. Building robust ML models should start with high-quality data, where training and real-world datasets are analyzed and curated to ensure their completeness, accuracy, and relevance in today's world. Failure to do so can compromise the reliability of a model's output.

Model Decay

A model's predictive ability is bound to degrade overtime, for example, in cases where new, evolving environments and data distributions are different from the historical data used to train the models. Huge differences in the data could impact the reliability of the system's predictions, leading to a *dataset shift*, which is the difference in the training and test data. Severe data shifts require model-correction techniques or retraining the model with more recent data to reestablish model accuracy. Another version of decay is *concept drift*, which occurs when data distributions stay the same, but the interpretation of the relationship between two or more features in the data changes. In this case, models produce results that are still accurate in their outputs, but are no longer relevant. To address model decay, it's important to monitor changes in model performance on a continual basis and utilize model correction techniques that allow for faster and less expensive recalibration of the model, rather than a complete model retraining on an updated dataset, which may be expensive to run.

Feature Stability

A *feature* is a variable used as an input in the ML systems. Features can be influenced by different variations over time. For example, a model that predicts housing price features could include house location, size, number of bedrooms, previous sale price, current valuation, and so on. Frequent variation of these features could impact a model's stability and performance, especially if relevant, unseen observations fall outside the range observed in training data. To avoid issues with feature stability, it's advisable to deliberately track changes and features as an indicator.

Precision versus Recall

Precision is a measure of a model's performance. Recall is the measure of completeness or quantity. There are often tradeoffs between precision and recall, as "high precision" indicates that a model returns more relevant results than irrelevant ones and "high recall" indicates that a model returns as many relevant results as are available. Due to this tradeoff, it's important to determine which balance is best for any given use case. Incorrect balances can impact the robustness of the model. In general, precision is important when a false positive could pose a critical problem. On the flip side, recall is important when a false negative could be a critical problem. Proper calibration of prevision versus recall should be carried out on each model on a constant basis.

Input Perturbations

As mentioned extensively, models can be tricked by deviations in input data, which can be exploited with malicious intent. Data poisoning is a type of attack on an ML model where the data used to train the model is intentionally contaminated to compromise the learning process. To avoid cases like these, it's important to have mitigation strategies in place, such as data sampling, which could include outliers (datapoints that are noticeably different from the rest), trends, and distributions. Building a *golden dataset* (a validated dataset with select cases from known sources, which already has a known, expected behavior) to ensure the integrity of the output is also another solution.

I hope you observed the dependencies between robustness, fairness, and privacy. Each of these areas is very important in running successful AI systems and are dependent on one another. It's impossible to have a responsible and ethical ML model if it addresses fairness issues alone, or privacy or robustness alone. Robustness is another aspect of the responsible AI framework that mustn't be ignored. It should form a part of the pretraining and fine-tuning processes in a consistent loop.

Summary

Overall, you've seen the importance in ensuring that your ML models are robust through the implementation of various techniques such as sampling, bias mitigation (at various levels), data balancing, and so on. This chapter covered different methodologies that can be applied in ML production to ensure your models are robust and secure from cyberattacks and malicious acts. This chapter also covered model challenges related to robustness. The next chapter, which is the final chapter of this book, reviews some of the challenges facing the ethics of AI and discusses how to address these.

PART III

Ethical Considerations

CHAPTER 10

AI Ethics

Once your ML models are robust and secure, not to mention all the other important responsible AI dimensions covered in this book, it's always a good idea to take a step back and use a wider lens to look at the overall challenges facing AI ethics. That way, you can learn how some of them can be averted, mitigated, and reduced. This final chapter looks at various ethical considerations and covers some of the most popular subfields in AI.

In December 2021, the AI industry, particularly the field of responsible AI, went through a major paradigm shift when Dr. Timnit Gebru and Dr. Emily Bender published a paper called "On the Dangers of Stochastic Parrots: Can Language Models Be Too Big?"[188] The publication stressed the need to pay more attention to the ethical implications of large-scale models, in particular language models, which have grown in popularity and adoption over the past few decades. Take ChatGPT for instance, a conversational chatbot launched in November 2022 by OpenAI. ChatGPT can write emails, essays, and web pages, generate answers to questions, create lines of code from prompts, debug and explain code, write resumes and cover letters, explain complex topics, solve math problems, and even give relationship advice. ChatGPT has become very popular, not only in tech domains but across all other sectors. It's already been adopted by major companies like Buzzfeed—a popular U.S. based digital media company, which laid off 12 percent of its workforce with the intention to use ChatGPT to create website content[189] (leading to a 200 percent stock increase[190]). Unsurprisingly, ChatGPT has provided false and inaccurate information like its predecessors. For example, it claimed that crushed porcelain could be added to breast milk to support the infant digestive system.[191] It also can provide convincing but wrong responses to coding questions, so much so that Stack Overflow (a question and answer website for programmers) had to temporarily ban its use.[192]

Apart from the imminent threat of AI replacing people's jobs, which is clearly seen from the Buzzfeed example, there are also many more threats that AI poses to human ethics that should not only be taken seriously but avoided as much as possible.

AI and ML models, in particular large-scale models, present environmental and financial risks due to the amount of computing process power required, which leads to a high demand for electricity. A separate paper by Emma Strubell and collaborators discovered that the carbon emissions and financial cost of large language models in particular have been skyrocketing since 2017, as these models need to be fed large amounts of data.[193]

Strubell's study also found that training a language model using a "neural architecture search" technique would produce the equivalent of 626,155 pounds of carbon dioxide, about the same amount as five average American cars or the total emissions of a 56 year old's entire life. Remembering this research was conducted in 2017, and many major advancements in the field have taken place since then, these figures would be so much higher today. To put this into perspective, ChatGPT currently uses 175 billion parameters. That's a lot of CO_2.

While you might be thinking that you don't need AI at all if it emits large amounts of carbon dioxide, significantly contributing to the decline of the planet, it's important to remember that AI is also solving the world's largest problems, including climate change, and has made many breakthroughs in healthcare. It has become a necessary and important part of technology that we can't live without. That said, it's critical that organizations and businesses that work on large-scale models and AI technology in general are not only aware of its dangers and its impact on the climate, but also take the necessary measures to reduce the impact.

For example, working on the efficiency of large language models through *quantization*—a process where data is converted to smaller precisions while performing all critical operations—could enable efficiency gains such as disk or memory savings and reduce *inference latency*[194] (the processing time it takes to process one unit of data), which will ultimately reduce the carbon footprint.

Ethical Considerations for Large Language Models

In addition to the issues mentioned in the last section, there are also some other ethical considerations to think about when building ML models/applications.

Prevalent Discriminatory Language in LLMs

Also mentioned in Gebru's paper is the risk of continued racist language. Language shifts due to social change, for example the MeToo and Black Lives Matter movements, aren't represented in AI models taken from the Internet, as these models won't produce or interpret these new cultural norms.

Another important and subtle problem is the failure to capture the language and norms of the countries and cultures of people who have less/no access to the Internet and so have a smaller online linguistic footprint. This means we'll continue to see AI-generated language reflecting the practices of the West—richer countries and communities. To overcome these challenges, there is the strong need for accountability and documentation of AI models, with a particular focus on large-scale models.

Working with Crowdworkers

Annotation or labeling by crowdworkers is an important and critical part of ML development. As crowdworker organizations manage this process, it's important that they following these guidelines:

- Protect the identity and privacy of the workers.

- Provide fair wages to the crowdworkers; this should be nothing less than the minimum wage of their residing countries.

- Limit their exposure to less toxic content if labeling language models, because these models are often annotated to identify toxic comments.

- If placing raters in specialized pools, it's important to be open and transparent about the processes being taken throughout the annotation process, including gaining rater consent if conducting a research study.

Inequality and Job Quality

Advances in language models and technologies due to the automation of tasks, such as ChatGPT, could potentially lead to job displacement and an increase in low-income jobs. For example moderating the content of a chatbot using a language model, which is of greater concern as it's anticipated that the full displacement of jobs by AI in the near future is seen as a lower risk, compared to the risk of the increase of low-income jobs.

There are also implications to job quality that could affect an employee's well-being. For example, it's been observed that although the use of robots in factories and warehouses has reduced some of the safety risks to employees and taken over some mundane tasks, workers have also seen an increase in the pace of work, as well as reductions in autonomy, human contact, and collaboration.[195] There's also the risk that individuals working with language applications could face similar challenges, for example, seeing an increase in monotonous tasks of monitoring and validating language outputs when working in customer service industries. This will invariably increase the pace of work and reduce human connection and autonomy.

Impact on Creatives

With the rise of large-scale models and their ability to generate art and language outputs, such as ChatGPT and DALL-E, it's quite clear that generative models are having a major impact on creative economies. They have the potential to undermine the profitability of creative industries, without necessarily violating copyright laws. Generating content that serves as a credible substitute for a type of human creativity that may have been protected by copyright means such work could be replaced. Further conversations and studies need to be conducted in this area, to understand the implications of generative models to the creative industry, and measures should be taken to upskill and prepare creatives for further AI advancements.

Disparate Access to Language Model Benefits

The benefits of language models, such as language-based productivity tools like personal virtual assistants, will be inaccessible to people with little to no Internet access or the required hardware. This means there's an uneven distribution of risks and benefits of language technologies, increasing the divide between wealthier and more advantaged groups, which further exacerbates economic inequalities. The increase in equality means that the single biggest driver influencing global inequality is technology, which should be prevented.

Ethical Considerations for Generative Models

Let's consider some additional ethical considerations for generative models in general. If you're new to the AI party and you're wondering what in the world *generative* AI means, here's a quick and simple definition—generative AI is a form of AI that can produce various types of functionalities in a single model, ranging from text, audio, image, video, and speech. Remember the various large-scale models covered in the previous chapters, such as Galactica, Tay, Dall-E, Stable Diffusion, Imagen, Parti, and so on? These are all generative models, and generative models in most recent times are increasingly wowing scientific communities, businesses, and the world at large with astounding capabilities, including LLMs. Sometimes they even produce outputs they were not even taught or trained to do, currently termed "emergent properties." The increasing progress made in this area shows that the race toward Artificial General Intelligence (an AI system that has full human cognitive capabilities) is still on. While we're still wondering if at some point we'll cohabit with robots who will act a human assistants (I hope I didn't strike a nerve!), it's advisable to be aware of the current ethical challenges of these incredibly interesting and ever-changing models.

Generative AI has been utilized across many domains. In finance, it's used for fraud detection and risk management; in education, it helps personalize student lessons; in healthcare, it's utilized to improve medical imagery and help with drug discovery.[196] Most of the ethical considerations around generative AI have been covered throughout the book, but there are a few additional areas to consider, discussed next.

Deepfake Generation

Deepfakes are a form of synthetic media, including images, videos, and audio, that have been digitally altered so that they appear to be someone else or doing something else. These are typically used maliciously or to spread false information Deepfakes are almost impossible to distinguish from real media. They can lead to numerous grave ethical implications such as the spread of disinformation and manipulation of public opinion. For example, many deepfake videos show political candidates in a bad light, saying or doing something they never said or did. There have been various videos/images of this nature on the Internet in recent times.

Another ethical concern is the ability of deepfakes to defame or harass individuals by distributing fake images and videos of them. This was witnessed in an app called Sensity that stripped women of their clothing using deepfake technology. It's important that the general public become AI literate and are able to distinguish deepfakes from original media and report such cases depending on the context. In the case of deepfake technology, it's crucial that people are educated on how to identify a deepfake. Deepfakes tend to display unusual skin tones, strange lighting, and oddly positioned shadows, which indicate the media could be fake. Unnatural eye movements, audio quality that might display poor lip-sync, odd word pronunciation, and robotic voices are other indications of deepfakes. Additional signs can be strange body shapes or movement, artificial facial expressions, and awkward postures or physiques.[197]

Truthfulness, Accuracy, and Hallucinations

This pertains specifically to LLMs, which are prone to provide incorrect answers with conviction and make up information. Do you remember the man from Belgium who recently took his life due to a chatbot convincing him that doing so would save the world from climate change, discussed in Chapter 3? LLMs (which are generative models) are known to fabricate stories and come up with conspiracy theories that could cause great harm. Although a lot of progress has been made in this area, where ChatGPT for example provides better answers than GPT-3, there are still ways that people have circumvented the guardrails placed on these LLMs to produce harmful responses for malicious purposes or adversarial testing. In fact, according to the TruthfulQA benchmark developed by Stanford University, "most generative models are truthful only 25 percent of the time."[198]

To address this issue, the AI research community is looking at new and emerging ways to produce less harmful outputs of generative models. One of the most recent topics is "AI alignment," where AI systems are developed to achieve the desired and intended goals. Some researchers are working with value-based systems, where they're working to ensure human values are aligned with the AI's functionalities. Another methodology being adopted in the industry is "controllability" using synthetic data. This method helps developers/researchers create synthetic data that they use to train the ML model on a given output, and rather than using instruction-tuning on the model, they utilize "prompt-tuning" methods.

The term "prompt-tuning" is slowly taking over the known methodology of instruction-tuning when it comes to generative models. Prompt-tuning is a more advanced method, whereby experienced linguists evaluate the output of an LLM and provide a few factual and correct prompts that the model is later trained on. Instruction-tuning is a traditional way to train language models by training a model with instructions and fine-tuning the model based on their performance and results. I'm optimistic that these various methods will provide more controlled results. As this is still a very new and fast-moving space, it's a bit premature and early to say issues with generative models will be solved anytime soon. In the meantime, an increase in AI literacy is needed across the world to educate everyone about the possibilities and potential harms of these technologies.

Copyright Infringement

This is increasingly becoming a major concern, especially in the creative industries. Copyright infringement is a major challenge with generative models. There have been several questions regarding ownership of work generated by the AI system, in terms of the application of copyright laws if produced by an AI, and using it for training purposes. Here's a funny story. Recently, a photographer won the prize for best photography at Sony's World Photography competition. After winning the award, the photographer admitted that he had been a "cheeky monkey" and that the photograph had actually been generated by AI. He claimed he wanted to raise awareness of the issues AI-generated art poses to the creative industries. Although the "cheeky monkey" won the award and refused it based on the fact that he didn't create the photograph, he brings up a very valid point. It was, and still is, very hard to detect art and sometimes content generated by AI. Today, these problems remain unresolved and some organizations have even banned the use of AI-generated images for the time being. For example, Getty Images has banned the "upload and sale of illustrations generated from AI art tools."[199]

Ethical Considerations for Computer Vision

While most AI models are headed toward generative AI, there are still singular AI modalities that exist and will continue to exist in the future, such as computer vision, which is widely used across smartphones, medical imagery, criminal justice agencies, autonomous vehicles, and so on. This section looks at a few of the ethical concerns for ML models using computer vision in addition to bias, which is a common denominator across various AI systems.

Issues of Fraud

According to the *Wall Street Journal,* between June 2020 and January 2021, there were over 80,000 attempts to hack into government facial recognition systems to claim unemployment benefits.[200] Hackers and fraudsters constantly use masks and photos to try to outsmart facial recognition technology in order gain entry to sites with someone else's identity.[201] Although the 80,000 attempts mentioned by the *Wall Street Journal* were during the COVID-19 pandemic, which saw an increase in fraudulent activity in general, fraud is a common problem in the computer vision field. This again highlights the need for heightened security against cybersecurity attacks.

Inaccuracies

The problem of inaccurate information is predominant in healthcare and disease detection, where noise can lead to inaccuracies and faulty diagnoses. For example, a computer vision system made false medical diagnoses based on the type of X-ray used, because it made a "false correlation between portable x-ray machines and a specific disease." It's important to constantly test the model for false negatives and positives in an attempt to determine the accuracy of the results.

Consent Violations

Violations of consent span across legal and ethical considerations. Facial recognition has been illegally used to collect personal data without consent, especially in the private sector. Remember the example of Clearview AI discussed in Chapter 2? Another relevant example is Apple's plan in 2021 to detect Child Sexual Abuse Material (CSAM) by scanning photos uploaded to iCloud by users. This controversial and privacy-infringing initiative led to public criticism and outcry, especially from privacy and security researchers, and digital rights groups on the concerns of undermining the privacy of Apple's users. Due to this outcry, Apple scrapped its plans a month later.[202] These consent violations infringe on many existing privacy laws. Also quite close to home are the large datasets of facial images that are used without user content. Some of these are not only used for commercial purposes but also to improve military and surveillance algorithms. It's been reported that personal data posted on the Internet is used as training data in surveillance applications across the world. To address this particular issue, you must ensure that data preserving practices, data ethics, and data curation techniques are in place. These have been discussed extensively in previous chapters of the book.

Summary

It's crucial to understand the various ethical implications of AI models on society and take the necessary steps to reduce these risks/harms. From identified harms related to inequality, low income jobs/job losses, increased carbon emissions, to low wages and the impact on creative communities, just to mention a few, it's important to consider the deleterious effects of AI/ML algorithms and products during the development and deployment phases. The stakes are high. The impact is huge, and in many cases irreparable/irreversible once deployed. The onus is on everyone in the field working on these systems to consider ethics and responsibility in the overall design of these technologies. The driver should go beyond profit-making and should include ethical considerations about the impact these systems have in the world in which we live.

APPENDIX A

References

The references in this appendix are numbered in line with the chapters in the book. Review these for further reference.

Chapter 1

1. Cambridge Dictionary, *Responsibility*, www.dictionary.cambridge.org/dictionary/english/responsibility.

2. European Commission, *Shaping Europe's Digital Future*, digital-strategy.ec.europa.eu/en/policies/european-approach-artificial-intelligence.

3. Livescience, *A Brief History of Artificial Intelligence*, www.livescience.com/49007-history-of-artificial-intelligence.

4. Blameless, *What Are Blameless Postmortems? (Do They Work? How?)*, www.blameless.com/blog/what-are-blameless-postmortems-do-they-work-how.

5. Wikipedia, *Trolley Problem*, en.wikipedia.org/wiki/Trolley_problem.

© Toju Duke 2023
T. Duke, *Building Responsible AI Algorithms*,
https://doi.org/10.1007/978-1-4842-9306-5

6. MIT Technology Review, *Should a Self-Driving Car Kill the Baby or the Grandma? Depends on Where You're From,* www.technologyreview.com/2018/10/24/139313/a-global-ethics-study-aims-to-help-ai-solve-the-self-driving-trolley-problem.

7. OECD.AI, *Accountability,* www.oecd.ai/en/dashboards/ai-principles/P9.

8. Analytics Insight, *Top Tech Companies Applying Artificial Intelligence for Social Good,* www.analyticsinsight.net/top-tech-companies-applying-artificial-intelligence-for-social-good/.

9. Grand View Research, *Artificial Intelligence Market Size, Share & Trends Analysis Report By Solution, By Technology (Deep Learning, Machine Learning), By End-use, By Region, and Segment Forecasts, 2023 - 2030,* www.grandviewresearch.com/industry-analysis/artificial-intelligence-ai-market.

10. *The New York Times, Wrongfully Accused by an Algorithm,* www.nytimes.com/2020/06/24/technology/facial-recognition-arrest.

11. County of Wayne, *WCPO Statement in Response to The New York Times Article Wrongfully Accused by an Algorithm,* June 24, 2020, int.nyt.com/data/documenthelper/7046-facial-recognition-arrest/5a6d6d0047295fad363b/optimized/full.pdf#page=1.

12. Womble Bond Dickinson, *Facial Recognition: A New Trend in State Regulation,* www.womblebonddickinson.com/us/insights/alerts/facial-recognition-new-trend-state-regulation.

13. Reuters, *U.S. Cities are Backing Off Banning Facial Recognition as Crime Rises,* www.reuters.com/world/us/us-cities-are-backing-off-banning-facial-recognition-crime-rises-2022-05-12/.

14. CNBC, *Rules Around Facial Recognition and Policing Remain Blurry,* www.cnbc.com/2021/06/12/a-year-later-tech-companies-calls-to-regulate-facial-recognition-met-with-little-progress.html.

15. Analytics Insight, *Which Countries Allow and which Ban AI Facial Recognition?* www.analyticsinsight.net/countries-allow-ban-ai-facial-recognition/.

16. European Parliament, *Regulating Facial Recognition in the EU,* www.europarl.europa.eu/RegData/etudes/IDAN/2021/698021/EPRS_IDA(2021)698021_EN.pdf.

17. Meta AI, *BlenderBot 3: A 175B Parameter, Publicly Available Chatbot that Improves its Skills and Safety Over Time,* ai.facebook.com/blog/blenderbot-3-a-175b-parameter-publicly-available-chatbot-that-improves-its-skills-and-safety-over-time.

18. Meta, *BlenderBot 3: An AI Chatbot That Improves Through Conversation,* www.about.fb.com/news/2022/08/blenderbot-ai-chatbot-improves-through-conversation.

19. Rockcontent, *40 Places to Find Open Data on the Web,* www.rockcontent.com/blog/data-sources.

20. Vice, Facebook's AI Chatbot: *Since Deleting Facebook My Life Has Been Much Better,* www.vice.com/en/article/qjkkgm/facebooks-ai-chatbot-since-deleting-facebook-my-life-has-been-much-better.

21. Techopedia, *Garbage In, Garbage Out (GIGO),* www.techopedia.com/definition/3801/garbage-in-garbage-out-gigo.

22. The Guardian, *Tay, Microsoft's AI Chatbot, Gets a Crash Course in Racism from Twitter,* www.theguardian.com/technology/2016/mar/24/tay-microsofts-ai-chatbot-gets-a-crash-course-in-racism-from-twitter.

23. The Verge, *Twitter Taught Microsoft's AI Chatbot to Be a Racist Asshole in Less Than a Day,* www.theverge.com/2016/3/24/11297050/tay-microsoft-chatbot-racist.

24. The London School of Economics and Political Science, *"F**k the Algorithm"?: What the World Can Learn from the UK's A-Level Grading Fiasco,* blogs.lse.ac.uk/impactofsocialsciences/2020/08/26/fk-the-algorithm-what-the-world-can-learn-from-the-uks-a-level-grading-fiasco.

25.   IAPP, *Digital Welfare Fraud Detection and the Dutch SyRI Judgment,* iapp.org/news/a/digital-welfare-fraud-detection-and-the-dutch-syri-judgment.

26.   UK Council for Internet Safety, *Child Safety Online: A Practical Guide for Parents and Carers Whose Children Are Using Social Media,* www.gov.uk/government/publications/child-safety-online-a-practical-guide-for-parents-and-carers/child-safety-online-a-practical-guide-for-parents-and-carers-whose-children-are-using-social-media.

Chapter 2

27.   Forrester, *Five AI Principles To Put In Practice,* www.forrester.com/blogs/five-ai-principles-to-put-in-practice.

28.   Haas School of Business, *What Does "Fairness" Mean for Machine Learning Systems?* haas.berkeley.edu/wp-content/uploads/What-is-fairness_-EGAL2.pdf.

29.   Arvind Narayanan, *21 Fairness Definitions and Their Politics,* fairmlbook.org/tutorial2.html.

30.   Google AI, *AI Principles 2020 Progress Update,* ai.google/static/documents/ai-principles-2020-progress-update.pdf.

31.   Google AI, *Artificial Intelligence at Google: Our Principles,* ai.google/principles.

32. U.S. Department of State, *The Organization for Economic Cooperation and Development (OECD)*, www.state.gov/the-organization-for-economic-co-operation-and-development-oecd.

33. OECD.AI, *Human-Centred Values and Fairness (Principle 1.2)*, oecd.ai/en/dashboards/ai-principles/P6.

34. Gabriel, I. *Artificial Intelligence, Values, and Alignment, Minds & Machines* 30, 411–437 (2020), doi.org/10.1007/s11023-020-09539-2.

35. Australian Government, Department of Industry, Science and Resources, *Australia's AI Ethics Principles,* www.industry.gov.au/data-and-publications/australias-artificial-intelligence-ethics-framework/australias-ai-ethics-principles.

36. BBC, *Google Apologises for Photos App's Racist Blunder,* www.bbc.co.uk/news/technology-33347866.

37. ProPublica, *How We Analyzed the COMPAS Recidivism Algorithm,* www.propublica.org/article/how-we-analyzed-the-compas-recidivism-algorithm.

38. Forbes, *Addressing AI's Biggest Problem: Trust,* www.forbes.com/sites/forbestechcouncil/2021/10/25/addressing-ais-biggest-problem-trust.

39. John Zerilli, Umang Bhatt, Adrian Weller, *How Transparency Modulates Trust in Artificial Intelligence*, Patterns, Volume 3, Issue 4, (2022), doi.org/10.1016/j.patter.2022.100455.

40. IBM, *Our Fundamental Properties for Trustworthy AI*, www.ibm.com/artificial-intelligence/ai-ethics-focus-areas.

41. The Alan Turing Institute, *Understanding Artificial Intelligence Ethics and Safety*, www.turing.ac.uk/sites/default/files/2019-06/understanding_artificial_intelligence_ethics_and_safety.pdf.

42. Virginia Dignum, *The ART of AI — Accountability, Responsibility, Transparency* (Blog), March 4, 2018, medium.com/@virginiadignum/the-art-of-ai-accountability-responsibility-transparency-48666ec92ea5.

43. Harvard Business Review, *How to Build Accountability Into Your AI*, hbr.org/2021/08/how-to-build-accountability-into-your-ai.

44. OECD.AI, *Accountability*, www.oecd.ai/en/dashboards/ai-principles/P9.

45. McKinsey Global Institute, *Applying Artificial Intelligence for Social Good*, www.mckinsey.com/featured-insights/artificial-intelligence/applying-artificial-intelligence-for-social-good.

46. Bernard Marr, *10 Wonderful Examples Of Using Artificial Intelligence (AI) For Good,* bernardmarr.com/10-wonderful-examples-of-using-artificial-intelligence-ai-for-good.

47. TechRepublic, *How AI Is Being Used for COVID-19 Vaccine Creation and Distribution,* www.techrepublic.com/article/how-ai-is-being-used-for-covid-19-vaccine-creation-and-distribution.

48. Newscientist, *DeepMind's Protein-Folding AI Cracks Biology's Biggest Problem,* www.newscientist.com/article/2330866-deepminds-protein-folding-ai-cracks-biologys-biggest-problem.

49. DeepMind, *AlphaFold Protein Structure Database,* alphafold.ebi.ac.uk.

50. United Nations, Department of Economic and Social Affairs, *World Population Prospects 2022,* www.un.org.development.desa.pd/files/wpp2022_summary_of_results.pdf.

51. International Food Policy Research Institute (IFPRI), *Global Nutrition Report 2016: From Promise to Impact: Ending Malnutrition by 2030,* www.ifpri.org/publication/global-nutrition-report-2016-promise-impact-ending-malnutrition-2030.

52. Global Food Security, *The Challenge,* www.foodsecurity.ac.uk/challenge.

53. United Nations, Economic and Social Council,
     *The role of Artificial Intelligence in achieving food
     security in the post Covid-19 era: Investing in a
     safe, nutritious and climate-resilient food systems,*
     www.un.org.ecosoc/files/files/en/2021doc/
     Statement%20Role%20of%20Artificial%20
     Intelligence%20in%20Food%20Security.pdf.

54. Food and Agriculture Organisation of the
     United Nations, *Can Artificial Intelligence Help
     Improve Agricultural Productivity?* www.fao.
     org/e-agriculture/news/can-artificial-
     intelligence-help-improve-agricultural-
     productivity.

55. Wired, *Phone-Powered AI Spots Sick Plants with
     Remarkable Accuracy,* www.wired.com/story/
     plant-ai.

56. Harvard Business School, Digital Initiative,
     *John Deere Subsidiary Blue River Technology
     is Transforming Agriculture Through Machine
     Learning,* d3.harvard.edu/platform-digit/
     submission/john-deere-subsidiary-blue-river-
     technology-is-transforming-agriculture-
     through-machine-learning.

57. AI for Good, *Robotics and AI to Predict and Fight
     Wildfires,* aiforgood.itu.int/robotics-and-ai-
     to-predict-and-fight-wildfires.

58. AI for Good, *7 AI Innovations Helping to Combat
     Climate Change,* aiforgood.itu.int/7-ai-
     innovations-helping-to-combat-climate-change.

59. IUCN, *Marine Plastic Pollution* www.iucn.org/
     resources/issues-brief/marine-plastic-
     pollution#:~:text=Impacts%20on%20marine%20
     ecosystems&text=Marine%20wildlife%20
     such%20as%20seabirds,to%20swim%2C%20and%20
     internal%20injuries.

60. The Ocean Cleanup, *Using AI to Monitor Plastic
     Density in the Ocean,* theoceancleanup.com/
     updates/using-artificial-intelligence-to-
     monitor-plastic-density-in-the-ocean.

61. Google Keyword, *Creating New Tree Shade with
     the Power of AI and Aerial Imagery,* blog.google/
     products/earth/helping-cities-seed-new-
     trees-with-tree-canopy-lab.

62. Google Keyword, *Expanding Our ML-Based
     Flood Forecasting,* blog.google/technology/ai/
     expanding-our-ml-based-flood-forecasting.

63. Microsoft AI, *AI for Humanitarian Action,*
     www.microsoft.com/en-us/ai/ai-for-
     humanitarian-action.

64. Microsoft AI, *AI for Humanitarian Action
     Projects,* www.microsoft.com/en-us/ai/ai-for-
     humanitarian-action-projects?activetab=pivot
     1%3aprimaryr3.

65. Unesco, *Draft Text of the Recommendation on the
     Ethics of Artificial Intelligence,* unesdoc.unesco.
     org/ark:/48223/pf0000377897.

66. European Commission, Directorate-General
     for Communications Networks, Content and
     Technology, *Ethics Guidelines for Trustworthy
     AI*, Publications Office (2019), `data.europa.eu/
     doi/10.2759/346720`.

67. Endpoint Protector, *EU vs US: What Are the
     Differences Between Their Data Privacy Laws?*
     `www.endpointprotector.com/blog/eu-vs-us-
     what-are-the-differences-between-their-data-
     privacy-laws`.

68. Enzuzo, *51 Biggest GDPR Fines and Penalties So Far*,
     `www.enzuzo.com/blog/biggest-gdpr-fines`.

69. Politico, *Amazon Fined €746M for Violating Privacy
     Rules*, `www.politico.eu/article/amazon-fined-
     e746m-for-violating-privacy-rules`.

70. NIST Computer Security Resource Center, *NIST
     Risk Management Framework*, `csrc.nist.gov/
     projects/risk-management/fisma-background`.

71. ICO. Information Commissioner's Office, *ICO
     Fines Facial Recognition Database Company
     Clearview AI Inc More than £7.5M and Orders UK
     Data to be Deleted*, `ico.org.uk/about-the-ico/
     media-centre/news-and-blogs/2022/05/ico-
     fines-facial-recognition-database-company-
     clearview-ai-inc`.

Chapter 3

72. VentureBeat, *MIT Takes Down 80 Million Tiny Images Dataset Due to Racist and Offensive Content,* venturebeat.com/ai/mit-takes-down-80-million-tiny-images-data-set-due-to-racist-and-offensive-content.

73. Silicon Republic, *80m Images Used to Train AI Pulled After Researchers Find String of Racist Terms,* www.siliconrepublic.com/machines/mit-database-racist-misogynist-discovery-abeba-birhane.

74. Antonio Torralba, Rob Fergus, Bill Freeman, *80 Million Tiny Images,* groups.csail.mit.edu/vision/TinyImages.

75. Mathematical Association of America, *Mathematical Treasure: Ishango Bone,* www.maa.org/press/periodicals/convergence/mathematical-treasure-ishango-bone.

76. ThinkAutomation, *The History of Data,* www.thinkautomation.com/histories/the-history-of-data.

77. United States Census Bureau, *The Hollerith Machine,* www.census.gov/history/www/innovations/technology/the_hollerith_tabulator.html.

78. ThinkAutomation, *The Data Lifecycle: Explained,* www.thinkautomation.com/eli5/the-data-lifecycle-explained.

79. The Open Data Institute, *Data Ethics Canvas,*
     theodi.org/wp-content/uploads/2019/07/ODI-
     Data-Ethics-Canvas-2019-05.pdf.

80. Harvard Business School Online, *5 Principles of
     Data Ethics for Business,* online.hbs.edu/blog/
     post/data-ethics.

81. Shift Paradigm, *Four Principles of Ethical Data
     Use,* www.shiftparadigm.com/insights/four-
     principles-of-ethical-data-use.

82. Dataethics, *Data Ethics Principles,* dataethics.eu/
     data-ethics-principles.

83. Thomson Reuters Practical Law, *Disparate
     Impact,* uk.practicallaw.thomsonreuters.
     com/1-502-2874.

84. Ethics of Data Curation, *Ethics in Data Curation,*
     ethicsofdatacuration.wordpress.com/ethics-
     in-data-curation.

85. American Psychological Association, *The
     Cognitive Bases of Anthropomorphism: From
     Relatedness to Empathy,* psycnet.apa.org/
     record/2015-01495-001.

86. Arleen Salles, Kathinka Evers & Michele
     Farisco, *Anthropomorphism in AI,* AJOB
     Neuroscience (2020), 11:2, 88-95, DOI:
     10.1080/21507740.2020.1740350.

87. National Library of Medicine, *Anthropomorphizing
     Without Social Cues Requires the Basolateral
     Amygdala,* pubmed.ncbi.nlm.nih.gov/30562137.

88. Euronews, *Man Ends His Life After an AI Chatbot 'Encouraged' Him to Sacrifice Himself to Stop Climate Change,* www.euronews.com/next/2023/03/31/man-ends-his-life-after-an-ai-chatbot-encouraged-him-to-sacrifice-himself-to-stop-climate.

89. Datagen, *Guide: Synthetic Data Generation,* datagen.tech/guides/synthetic-data/synthetic-data-generation.

90. Mahima Pushkarna, Andrew Zaldivar, Oddur Kjartansson, *Data Cards: Purposeful and Transparent Documentation for Responsible AI,* doi.org/10.48550/arXiv.2204.01075.

91. Timnit Gebru, Jamie Morgenstern, Briana Vecchione, Jennifer Wortman Vaughan, Hanna Wallach, Hal Daumé III, Kate Crawford, *Datasheets for Datasets,* Communications of the ACM (2021), doi.org/10.48550/arXiv.1803.09010.

92. Mitchell, Margaret, Wu, Simone, Zaldivar, Andrew, Barnes, Parker Vasserman, Lucy, Hutchinson, Ben, Spitzer, Elena, Raji, Inioluwa Deborah, and Gebru, Timnit, *Model Cards for Model Reporting,* Association for Computing Machinery (2019), doi.org/10.1145/3287560.3287596.

93. The Open Data Institute, *Updating the Data Ethics Canvas,* theodi.org/article/updating-the-data-ethics-canvas.

94. People + AI Research, *Know Your Data: A New Tool to Explore Datasets,* medium.com/people-ai-research/know-your-data-a-new-tool-to-explore-datasets-ba45b7665695.

95. Yuki M. Asano, Christian Rupprecht, Andrew Zisserman, and Andrea Vedaldi, *PASS: An ImageNet Replacement for Self-Supervised Pretraining Without Humans,* NeurIPS Track on Datasets and Benchmarks (2021), doi.org/10.48550/arXiv.2109.13228.

96. Parul Pandey, *Overcoming ImageNet Dataset Biases with PASS* (Blog), October 2021, towardsdatascience.com/overcoming-imagenet-dataset-biases-with-pass-6e54c66e77a.

Chapter 4

97. Reuters, *Amazon Scraps Secret AI Recruiting Tool That Showed Bias Against Women,* www.reuters.com/article/us-amazon-com-jobs-automation-insight-idUSKCN1MK08G.

98. Oxford Reference, *Pseudoscience,* www.oxfordreference.com/view/10.1093/acref/9780199594009.001.0001/acref-9780199594009-e-1007.

99. BBC, *AI Tools Fail to Reduce Recruitment Bias - Study,* www.bbc.co.uk/news/technology-63228466.

100.   Mehrabi, Ninareh, Morstatter, Fred, Saxena,
        Nripsuta, Lerman, Kristina, and Galstyan, Aram,
        A Survey on Bias and Fairness in Machine Learning,
        (2019), doi.org/10.48550/arxiv.1908.09635.

101.   Arvind Narayanan, *TL;DS 21 Fairness Definition
        and Their Politics by Arvind Narayanan* (Blog),
        September 2019, shubhamjain0594.github.io/
        post/tlds-arvind-fairness-definitions.

102.   Carnegie Mellon University, Statistics &
        Data Science, *Classifier Fairness,* www.stat.
        cmu.edu/~cshalizi/dm/20/lectures/25/
        lecture-25.html.

103.   Conor O'Sullivan, *Analysing Fairness in Machine
        Learning (with Python),* (Blog), April 2022,
        towardsdatascience.com/analysing-fairness-
        in-machine-learning-with-python-96a9ab0d0705.

104.   Center for Critical Race + Digital Studies, *Types
        of Bias,* www.criticalracedigitalstudies.com/
        peoples-guide-posts/sources-of-bias.

105.   Datatron, *How Gender Bias Led to the Scrutiny
        of the Apple Card,* datatron.com/how-gender-
        bias-led-to-the-scrutiny-of-the-apple-
        card/#:~:text=It%20was%20found%20that%20
        women,social%20security%20number%2C%20and%20
        birthdate.

106.   Cathy O'Neil, *Weapons of Math Destruction:
        How Big Data Increases Inequality and Threatens
        Democracy.*

107. Northpointe, *Practitioner's Guide to COMPAS Core,* s3.documentcloud.org/documents/2840784/ Practitioner-s-Guide-to-COMPAS-Core.pdf.

108. MIT Technology Review, *The New Lawsuit That Shows Facial Recognition Is Officially a Civil Rights Issue,* www.technologyreview. com/2021/04/14/1022676/robert-williams-facial-recognition-lawsuit-aclu-detroit-police.

109. IBM Research, *AI Fairness 360,* aif360. mybluemix.net.

110. People + AI Research, *Learning Interpretability Tool (LIT),* pair-code.github.io/lit.

111. GitHub, *Fairlearn,* github.com/fairlearn/ fairlearn.

112. PwC, *PwC's Responsible AI,* www.pwc.com/gx/ en/issues/data-and-analytics/artificial-intelligence/what-is-responsible-ai.html.

113. GitHub, *Audit-AI,* github.com/pymetrics/ audit-ai.

114. Synthesized, *FairLens Unveiled: Discover And Measure Data Bias,* www.synthesized.io/fairlens.

115. Fairgen, *Three Tools to Utilise Generative AI for Fairer Data,* www.fairgen.ai.

116. Microsoft, *Fairlearn: A Toolkit for Assessing and Improving Fairness in AI,* www.microsoft.com/en-us/research/uploads/prod/2020/05/Fairlearn_ WhitePaper-2020-09-22.pdf.

Chapter 5

117. MIT Technology Review, *Why Meta's Latest Large Language Model Survived Only Three Days Online,* www.technologyreview.com/2022/11/18/1063487/meta-large-language-model-ai-only-survived-three-days-gpt-3-science.

118. TNW, *Meta Takes New AI System Offline Because Twitter Users Are Mean,* thenextweb.com/news/meta-takes-new-ai-system-offline-because-twitter-users-mean.

119. The AI Journal, *AI Safety: What Does It Mean And What Do You Need To Know?* aijourn.com/ai-safety-what-does-it-mean-and-what-do-you-need-to-know.

120. Faculty, *What Is AI Safety?* faculty.ai/blog/what-is-ai-safety.

121. Alex Moltzau, *Avoiding Side Effects and Reward Hacking in Artificial Intelligence* (Blog), July 2019, towardsdatascience.com/avoiding-side-effects-and-reward-hacking-in-artificial-intelligence-18c28161190f.

122. Input, *This AI posted on 4chan for days before being unmasked,* www.inverse.com/input/tech/artificial-intelligence-4chan-bot.

123. Anil Tilbe, *Zero-shot vs Few-shot Learning: Key Insights with 2022 Updates* (Blog), June 2022, pub.towardsai.net/zero-shot-vs-few-shot-learning-50-key-insights-with-2022-updates-17b71e8a88c5.

124. Weidinger, Laura et al., *Taxonomy of Risks Posed by Language Models*, FAccT 2022. `doi.org/10.1145/3531146.3533088`.

125. Sunipa Dev, Emily Sheng, Jieyu Zhao, Aubrie Amstutz, Jiao Sun, Yu Hou, Mattie Sanseverino, Jiin Kim, Akihiro Nishi, Nanyun Peng, and Kai-Wei Chang, 2022. *On Measures of Biases and Harms in NLP*, Association for Computational Linguistics: AACL-IJCNLP, (2022), `aclanthology.org/2022.findings-aacl.24`.

126. Aylin Caliskan, Joanna J. Bryson, and Arvind Narayanan, *Semantics Derived Automatically from Language Corpora Contain Human-Like Biases*, American Association for the Advancement of Science (2017), `doi.org/10.1126%2Fscience.aal4230`.

127. Abid, Abubakar and Farooqi, Maheen and Zou, James, *Persistent Anti-Muslim Bias in Large Language Models* (2021), `doi.org/10.48550/arxiv.2101.05783`.

128. Inioluwa Deborah Raji, *Handle with Care: Lessons for Data Science from Black Female Scholars (2020)*, `doi.org/10.1016/j.patter.2020.100150`.

129. Hwang, Gilhwan, Lee, Jeewon, Oh, Cindy Yoonjung, and Lee, Joonhwan, *It Sounds Like A Woman: Exploring Gender Stereotypes in South Korean Voice Assistants*, Association for Computing Machinery (2019), `doi.org/10.1145/3290607.3312915`.

130. People + AI, *People + AI Guidebook,* design.google/
     aiguidebook.

131. Lynn H Kaack, Priya L Donti, Emma Strubell, George
     Kamiya, Felix Creutzig, et al., *Aligning Artificial
     Intelligence with Climate Change Mitigation* (2021),
     hal.archives-ouvertes.fr/hal-03368037.

132. Alexandre Georgieff & Anna Milanez, *What
     Happened to Jobs at High Risk of Automation?,*
     OECD Social, Employment and Migration
     Working Papers (2021), ideas.repec.org/p/oec/
     elsaab/255-en.html.

133. Rasa, *Benchmarking Language Models* (Blog),
     September 2022, rasa.com/blog/benchmarking-
     language-models.

134. Nadeem, Moin, Bethke, Anna, and Reddy, Siva,
     *StereoSet: Measuring Stereotypical Bias in Pretrained
     Language Models (2020),* doi.org/10.48550/
     arxiv.2004.09456.

135. Nozza, Debora, Bianchi, Federico, Hovy, Dirk,
     *{HONEST}: Measuring Hurtful Sentence Completion
     in Language Models,* Association for Computational
     Linguistics: Human Language Technologies (2021),
     aclanthology.org/2021.naacl-main.191.

136. GitHub, *Winogender-Schemas,* github.com/
     rudinger/winogender-schemas.

137. Papers with code, *OpenWebText,* paperswithcode.
     com/dataset/openwebtext.

138. Perspective, *Using Machine Learning to Reduce Toxicity Online,* perspectiveapi.com.

139. Open Review, *Appendix Overview,* openreview.net/attachment?id=u46CbCaLufp&name=supplementary_material.

140. Adversa, *AI Risk Management for Automotive Industry,* adversa.ai/ai-risk-management-automotive-industry.

Chapter 6

141. Faculty, *What is 'Human-in-the-Loop'? And Why Is It More Important Than Ever?* faculty.ai/blog/what-is-human-in-the-loop.

142. Vikram Singh Bisen, *What Is Human in the Loop Machine Learning: Why & How Used in AI?* (Blog), May 2020, medium.com/vsinghbisen/what-is-human-in-the-loop-machine-learning-why-how-used-in-ai-60c7b44eb2c0.

143. Clickworker, *Human in the Loop: The Human in the Machine,* www.clickworker.com/customer-blog/human-in-the-loop-ml.

144. Google Cloud, *Human-in-the-Loop Overview,* cloud.google.com/document-ai/docs/hitl.

145. Jigsaw, Daniel Borkan, Jeff Sorensen, and Lucy Vasserman, *Exploring the Role of Human Raters in Creating NLP Datasets* (Blog), November 2019, medium.com/jigsaw/creating-labeled-datasets-and-exploring-the-role-of-human-raters-56367b6db298.

146. Goyal, Nitesh, Kivlichan, Ian, Rosen, Rachel, and Vasserman, Lucy, *Is Your Toxicity My Toxicity? Exploring the Impact of Rater Identity on Toxicity Annotation* (2022), doi.org/10.48550/arxiv.2205.00501.

147. Sap, Maarten, Card, Dallas, Gabriel, Saadia, Choi, Yejin, and Smith, Noah A, *The Risk of Racial Bias in Hate Speech Detection* (2019), aclanthology.org/P19-1163.

148. Internetlab, *Drag Queens and Artificial Intelligence: should Computers Decide What Is 'Toxic' on the Internet?* internetlab.org.br/en/news/drag-queens-and-artificial-intelligence-should-computers-decide-what-is-toxic-on-the-internet.

149. Kaggle, *Jigsaw Specialized Rater Pools Dataset,* www.kaggle.com/datasets/google/jigsaw-specialized-rater-pools-dataset.

150. Kaggle, *Jigsaw Unintended Bias in Toxicity Classification,* www.kaggle.com/c/jigsaw-unintended-bias-in-toxicity-classification/data.

151. Patel, B.N., Rosenberg, L., Willcox, G. et al., *Human–machine partnership with artificial intelligence for chest radiograph diagnosis,* doi.org/10.1038/s41746-019-0189-7.

Chapter 7

152. MIT News, *Explained: Neural Networks,* news.mit.edu/2017/explained-neural-networks-deep-learning-0414.

153. Dataversity, *A Brief History of Machine Learning,* www.dataversity.net/a-brief-history-of-machine-learning.

154. IBM, *AI vs. Machine Learning vs. Deep Learning vs. Neural Networks: What's the Difference?* www.ibm.com/cloud/blog/ai-vs-machine-learning-vs-deep-learning-vs-neural-networks.

155. Mahima Pushkarna, Andrew Zaldivar, Oddur Kjartansson, *Data Cards: Purposeful and Transparent Documentation for Responsible AI,* doi.org/10.48550/arXiv.2204.01075.

156. Google Model Cards, *The Value of a Shared Understanding of AI Models,* modelcards.withgoogle.com/about.

157. IBM Research, *AI Explainability 360,* aix360.mybluemix.net.

158. Harvard Business Review, *How to Build Accountability into Your AI,* hbr.org/2021/08/how-to-build-accountability-into-your-ai.

159. Investopedia, *Game Theory,* www.investopedia.com/terms/g/gametheory.asp.

160. Lundberg, Scott and Lee, Su-In, *A Unified Approach to Interpreting Model Predictions* (2017), doi.org/10.48550/arxiv.1705.07874.

161. Advancing Analytics, *How to Explain Your Machine Learning Model Using SHAP?* www.advancinganalytics.co.uk/blog/2021/7/14/shap.

162. SHAP, *Welcome to the SHAP Documentation,* shap.readthedocs.io/en/latest.

163. KDnuggets, *Introduction to Local Interpretable Model-Agnostic Explanations (LIME),* www.kdnuggets.com/2016/08/introduction-local-interpretable-model-agnostic-explanations-lime.html.

164. Explainer Dashboard, *Summary,* explainerdashboard.readthedocs.io/en/latest.

165. PAIR Explorables, *Searching for Unintended Biases with Saliency,* pair.withgoogle.com/explorables/saliency.

Chapter 8

166. Galaxkey, *Japanese Comic Book Application Suffers Extensive Data Breach,* www.galaxkey.com/blog/japanese-comic-book-application-suffers-extensive-data-breach.

167. World Economic Forum, *Why Artificial Intelligence Design Must Prioritize Data Privacy,* www.weforum.org/agenda/2022/03/designing-artificial-intelligence-for-privacy.

168. Google Research, *Federated Learning: Collaborative Machine Learning without Centralized Training Data,* ai.googleblog.com/2017/04/federated-learning-collaborative.html.

169. People + AI Research, *How Federated Learning Protects Privacy,* pair.withgoogle.com/explorables/federated-learning.

170. ACMQueue, *Federated Learning and Privacy,* queue.acm.org/detail.cfm?id=3501293.

171. Harvard University Privacy Tools Project, *Differential Privacy,* privacytools.seas.harvard.edu/differential-privacy.

172. Analytic Steps, *What Is Differential Privacy and How Does It Work?* www.analyticssteps.com/blogs/what-differential-privacy-and-how-does-it-work.

173. Bagdasaryan, Eugene and Shmatikov, Vitaly, *Differential Privacy Has Disparate Impact on Model Accuracy* (2019), `doi.org/10.48550/ arxiv.1905.12101`.

174. Suriyakumar, Vinith M., Papernot, Nicolas, Goldenberg, Anna, and Ghassemi, Marzyeh, *Chasing Your Long Tails: Differentially Private Prediction in Health Care Settings* (2020), `doi. org/10.48550/arxiv.2010.06667`.

175. PAIR Explorables, *Can a Model Be Differentially Private and Fair?* `pair.withgoogle.com/ explorables/private-and-fair`.

176. Papernot, Nicolas, Thakurta, Abhradeep, Song, Shuang, and, Steve, and Erlingsson, Úlfar, *Tempered Sigmoid Activations for Deep Learning with Differential Privacy* (2020), `doi.org/10.48550/ arxiv.2007.14191`.

177. Lee, Katherine, Ippolito, Daphne, Nystrom, Andrew, Zhang, Chiyuan, Eck, Douglas, Callison-Burch, Chris, and Carlini, Nicholas, *Deduplicating Training Data Makes Language Models Better* (2021), `doi. org/10.48550/arxiv.2107.06499`.

Chapter 9

178. Vector Institute, *Machine Learning Robustness: New Challenges and Approaches,* `vectorinstitute. ai/2022/03/29/machine-learning-robustness- new-challenges-and-approaches`.

179. Chen, Hongge, *Robust Machine Learning Models and Their Applications,* Massachusetts Institute of Technology, dspace.mit.edu/handle/1721.1/130760.

180. Hendrycks, Dan, Zhao, Kevin, Basart, Steven, Steinhardt, Jacob, and Song, Dawn, *Natural Adversarial Examples,* (2019), doi.org/10.48550/arxiv.1907.07174.

181. Joseph Early, *Your Car May Not Know When to Stop — Adversarial Attacks Against Autonomous Vehicles* (Blog), September 2019, towardsdatascience.com/your-car-may-not-know-when-to-stop-adversarial-attacks-against-autonomous-vehicles-a16df91511f4.

182. Periculum, *Practical Ways To Improve The Robustness of Machine Learning Models,* www.periculum.io/post/machine-learning-models.

183. PAIR Explorables, *From Confidently Incorrect Models to Humble Ensembles,* pair.withgoogle.com/explorables/uncertainty-ood.

184. Scribbr, *Sampling Methods | Types, Techniques & Examples,* www.scribbr.com/methodology/sampling-methods.

185. Geeks for Geeks, *Hyperparameter Tuning,* www.geeksforgeeks.org/hyperparameter-tuning.

186. Towards Data Science, *Using Adversarial Debiasing to Reduce Model Bias,* towardsdatascience.com/reducing-bias-from-models-built-on-the-adult-dataset-using-adversarial-debiasing-330f2ef3a3b4.

187.   Vector Institute, *Machine Learning Robustness: New Challenges and Approaches,* `vectorinstitute.ai/2022/03/29/machine-learning-robustness-new-challenges-and-approaches.`

Chapter 10

188.   Bender, Emily M., Gebru, Timnit, McMillan-Major, Angelina, and Shmitchell, Shmargaret, *On the Dangers of Stochastic Parrots: Can Language Models Be Too Big?* 🦜, Association for Computing Machinery (2021), `doi.org/10.1145/3442188.3445922.`

189.   ITP.net, *Buzzfeed to Use ChatGPT for Content Creation After 12% Employee Layoff,* `www.itp.net/emergent-tech/buzzfeed-to-use-chatgpt-for-content-creation-after-12-employee-layoff.`

190.   Forbes, *BuzzFeed To Use ChatGPT's AI For Content Creation, Stock Up 200%+,* `www.forbes.com/sites/chriswestfall/2023/01/26/buzzfeed-to-use-chatgpts-ai-for-content-creation-stock-up-200/?sh=1efdea2b7eae.`

191.   Twitter, Gary Marcus, *Describe How Crushed Porcelain Added to breast Milk Can Support the infant Digestive System* (2022), `twitter.com/GaryMarcus/status/1599584310633594881.`

192.   ZDNET, *Stack Overflow Temporarily Bans Answers from OpenAI's ChatGPT Chatbot,* `www.zdnet.com/article/stack-overflow-temporarily-bans-answers-from-openais-chatgpt-chatbot.`

193. Strubell, Emma, Ganesh, Ananya, and McCallum, Andrew, *Energy and Policy Considerations for Deep Learning in NLP* (2021), doi.org/10.48550/arxiv.1906.02243.

194. OpenGenus IQ, *Basics of Quantization in Machine Learning (ML) for Beginners,* iq.opengenus.org/basics-of-quantization-in-ml.

195. UC Berkeley Center for Labor Research and Education, *The Future of Warehouse Work: Technological Change in the U.S. Logistics Industry,* laborcenter.berkeley.edu/pdf/2019/Future-of-Warehouse-Work.pdf.

196. AI Multiple, *Generative AI Ethics: Top 6 Concerns,* research.aimultiple.com/generative-ai-ethics.

197. Q5id, *How to Spot and Identify a Deepfake: 7 Proven Techniques,* q5id.com/blog/how-to-spot-and-identify-a-deepfake-7-proven-techniques#:~:text=Unusual%20skin%20tones%2C%20stains%2C%20strange,is%20a%20deepfake%20or%20not.

198. Stanford University, Human-Centered Artificial Intelligence, *Artificial Intelligence Index Report 2022,* aiindex.stanford.edu/wp-content/uploads/2022/03/2022-AI-Index-Report_Master.pdf.

199. The Verge, *Getty Images Bans AI-Generated Content Over Fears of Legal Challenges,* www.theverge.com/2022/9/21/23364696/getty-images-ai-ban-generated-artwork-illustration-copyright.

200. The National Law Review, *As Facial Recognition Technology Surges, Organizations Face Privacy and Cybersecurity Concerns, and Fraud,* www.natlawreview.com/article/facial-recognition-technology-surges-organizations-face-privacy-and-cybersecurity.

201. Innodata, *Ethical Issues in Computer Vision and Strategies for Success,* innodata.com/ethical-issues-in-computer-vision-and-strategies-for-success.

202. Wired, *Apple Kills Its Plan to Scan Your Photos for CSAM. Here's What's Next,* www.wired.com/story/apple-photo-scanning-csam-communication-safety-messages.

Index

A

Accountability, 6–9, 22, 23, 42, 82, 108, 111, 116, 139

Accountability dimensions
    data, 112
    explainable AI tools, 113–116
    governance structures, 112
    monitoring plans, 113
    performance goals and metrics, 112

Adversarial debiasing, 129

Aggregation bias, 68, 70

AI alignment, 144

AI chatbot, 10, 11, 35, 46–48

AI ethics, *see* Ethical considerations

AI Explainability 360, 110

AI framework, 3, 12, 15, 61, 77, 117, 134

AI harms
    discrimination, hate speech and exclusion, 81, 82
    environmental and social, 85, 86
    HCI, 84, 85
    information hazards, 82, 83
    malicious uses, 83
    misinformation harms, 83

    mitigations and technical considerations, 87–90
    types, 80

AI lifecycle, 22, 31, 111

AI modalities, 80, 86, 145

AI principles
    accountability, 22, 23
    Australian government, 18–20
    "black box" problem, 20
    development, 15
    fairness, bias and human-centered values, 15–17
    guidelines, 15
    organizations and governing bodies, 15
    privacy, 31–35
    safety, 31–35
    security, 31–35
    social benefits, 23–31
    transparency and trust, 20, 21

AI recruiting tool, 61

AI recruitment software, 61

AI safety
    autonomous learning agents, 78, 79
    challenges, 78
    definition, 78

© Toju Duke 2023
T. Duke, *Building Responsible AI Algorithms*,
https://doi.org/10.1007/978-1-4842-9306-5